TV Sound Operations

Glyn Alkin

FOCAL PRESS
London & New York

ISBN 0 240 50865 3 (Paperback)

Printed in Great Britain by Johnson Riddle & Co. Ltd. and bound by the Pitman Press

MEDIA MANUALS

TV CAMERA OPERATION
Gerald Millerson

BASIC TV STAGING
Gerald Millerson

THE USE OF MICROPHONES
Alec Nisbett

YOUR FILM & THE LAB
I. Bernard Happé

TV SOUND OPERATIONS
Glyn Alkin

TV LIGHTING METHODS
Gerald Millerson

16mm FILM CUTTING
John Burder

USING VIDEOTAPE
J. F. Robinson, & P. H. Beards

THE SMALL TV STUDIO
Alan Bermingham, Michael Talbot-Smith,
John Symons, & Ken Angold-Stephens.

Forthcoming titles

Sound Recording & Reproduction
The Rostrum Camera
Motion Picture Camera — Choice & Technique
The Lens in Action
The Animation Rostrum
The Lens & All Its Jobs
Basic Film Technique

Contents

INTRODUCTION

THE NATURE OF SOUND **10**
- Sound intensity
- Loudness
- The subjective nature of sound

NATURE OF TELEVISION SOUND **12**
- Relative importance of picture and sound
- Relative complexity of picture and sound
- Comparison between sound and light

THE TELEVISION SOUND SUPERVISOR **14**
- Sound balance and control
- Staff control

PRESSURE OPERATED MICROPHONES **16**
- Moving coil microphone
- Electrostatic (capacitor) microphone
- Electret diaphragms

PRESSURE GRADIENT MICROPHONES **18**
- Ribbon microphone
- Directional characteristics
- Unidirectional ribbon microphones
- Unidirectional electrostatic microphones

DOUBLE ELEMENT MICROPHONES **20**
- Combination of pressure and pressure-gradient types
- Back-to-back electrostatics

SUPER-DIRECTIONAL MICROPHONES **22**
- The reflector microphone
- The rifle microphone

STAND AND HAND-HELD MICROPHONES **24**
- Hand held microphones

LAVALIER MICROPHONES **26**
- Supplying necessary HF lift
- Miniature electrostatic microphones

RADIO MICROPHONES **28**
- Transmission systems
- Site surveys

FISHING RODS **30**
- Type of microphone
- Microphone suspension
- Operation

SLUNG MICROPHONES **32**
- Microphones in shot
- Method of suspension

HIDDEN MICROPHONES **34**
- Maintaining microphone characteristic
- Sound perspective

SOUND BOOMS **36**
- Operation of the boom
- The boom pram

BOOM MICROPHONES **38**
- Directional characteristic
- Acceptance angle
- Construction and mounting

THE SCOPE OF THE BOOM **40**
- Advantages
- Disadvantages

BOOM OPERATING **42**
- Operator's responsibilities

BASIC LIGHTING REQUIREMENTS **44**
- Lamp functions

THE BOOM AND LIGHTING **46**
- Multi key lights
- Heat problems
- Lamp sing

THE BOOM AND CAMERA MOVEMENT **48**
- Cable arrangements
- Boom positioning

PLANNING FOR THE BOOM **50**
- The studio plan
- The elevations

PRODUCTION TALKBACK ARRANGEMENTS **52**
- Director's talkback
- Floor manager's talkback
- Camera talkback

SOUND AND LIGHTING TALKBACK **54**
- Sound talkback
- Reverse talkback
- Lighting talkback
- Wall points

SOUND CONTROL **56**
- Basic requirements
- Use of loudspeakers
- The importance of listening levels

THE POINTS OF CONTROL **58**
- Pre-set (or balance) attenuator
- Channel fader
- Group fader
- Main fader

THE SOUND CONTROL DESK **60**
- Programme volume control
- Frequency response control
- Multiway clean feed

PROGRAMME METERS **62**

AUTOMATIC CONTROL **64**
- The limiter
- The compressor

FREQUENCY RESPONSE CONTROL **66**
- Filters
- Response selection amplifier
- The octave filter
- Notch filters

STUDIO ACOUSTICS **68**
- Reverberation time
- Decay curve
- Diffusion
- Television studio acoustics

THE ACOUSTIC EFFECT OF SCENERY **70**
- Acoustic planning
- Design consideration
- Acoustic effect of materials
- The importance of shape

ARTIFICIAL REVERBERATION **72**
- The echo room
- The reverberation plate
- Reverberation springs

AMBIOPHONY **74**
- Assisted resonance
- Ambiophony
- Reverberation
- Setting up procedure

TALKS AND LECTURES **76**
- Introductory music
- Microphone arrangements

INFORMAL DISCUSSIONS **78**

FORMAL DISCUSSIONS AND PANEL GAMES **80**

PLANNING FOR STUDIO AUDIENCES **82**
- Visibility
- Operator discipline
- Production routine

AUDIENCE REINFORCEMENT **84**
- High level system
- Low level system
- Compromise arrangement

PA CONNECTION POINTS **86**
- Before the fader PA
- After the fader PA
- Combined arrangement

AUDIENCE REACTION **88**
- Sound character
- Separation from PA
- Milking the audience

AUDIENCE PARTICIPATION **90**
- Set up arrangement
- Hand held microphones
- Rifle microphones
- Booms
- Parabolic microphones

DISCUSSIONS BETWEEN REMOTE CENTRES **92**
- Deaf aids
- Loudspeakers
- Line up routine

PLANNING FOR DRAMA **94**
- Filming or pre-recording
- Planning meeting

CO-ORDINATING SOUND AND PICTURE **96**
- Perspective
- Distance

DIALOGUE AND SPLIT SOUND **98**
- Split sound
- Up stage — down stage split

GROUPING PROBLEMS **100**
- Tables

HEIGHT DISCREPANCY **102**
- Long shot to upper room or balcony
- Action on rostra

CEILINGS, BEAMS AND WINDOWS **104**
- Windows
- Reverse angles

COLUMNS, ARCHES AND DOORWAYS **106**
- Arches
- Doorways

VEHICLES **108**
- railway carriages
- Lifts
- Motor cars

BACK PROJECTION AND CSO **110**
- Back projection
- Colour separation overlay

WEATHER AND SOUND PROBLEMS **112**
- Wind
- Wind machines
- Trick lines
- Rain

GUNSHOTS **114**
- Real guns
- Electronic effects
- Foldback

TELEPHONE AND INTERCOM EFFECTS **116**
Telephone distort effects
Effects switching
Intercoms
EXTERIOR SEQUENCES IN THE STUDIO **118**
Exterior scenes in the studio
Use of super-directional microphones
Footsteps
LOCATION SHOOTING ON FILM **120**
The synchronising methods
ELECTRONIC CAMERAS ON LOCATION **122**
Weather
Noise
Use of booms
Camera mounted microphones
Radio microphones
SOUND EFFECTS IN THE STUDIO **124**
Unreal 'props'
Off stage sounds
Bells
Telephones
GRAMOPHONE OPERATING **126**
Groove locating
Turntable mechanism
Finding cues
Effects discs
Slip mats
TAPE RECORDING **128**
Incidental music recording
TAPE EDITING **130**
Cutting the tape
Joining the tape
Leader tapes
TELEVISING THE PIANO **132**
Tone production
Energy distribution
Choice of microphone position
TELEVISING THE RHYTHM PIANO **134**
The piano in a dance band
Microphone techniques
Multitone (jangle box) pianos
Microphone positioning
VARIOUS KEYBOARD INSTRUMENTS **136**
Harpsichord
Clavichord
Celeste
Organ
STRINGS **138**
Violin
Viola
Cello
Double bass
Rostrum resonance
WOODWIND INSTRUMENTS **140**
Type of approach
Flute and piccolo
Oboe
Clarinet
Bassoon
BRASS INSTRUMENTS **142**
Trumpets and trombones
Mutes
Horns
Euphonium and tubas
CHAMBER ORCHESTRA **144**
Pictorial presentation
String quartet
Classical orchestra
SYMPHONY ORCHESTRA **146**
The factor of size
Television perspective
Orchestral balance for television
OPERA **148**
Studio opera
Separate orchestral studio
Theatre relays of opera
THE LIGHT ORCHESTRA **150**
Light music balance
Multi microphone techniques
DANCE BAND **152**
Rhythm section
Rhythm piano
Double bass
Drums
Guitar
VARIETY ORCHESTRA **154**
Out of shot orchestra
In shot orchestra
Vocal separation
CONCERT ORCHESTRA OR SHOWBAND **156**
String treatment
Dance band section
Visual considerations

POP GROUPS **158**
The 'folk' group
'Electronic' groups
Recording technique
Technical treatment
SOUND AND VTR **160**
Discontinuous recording
Types of VTR machine
Helical scan VTR
SOUND AND VT EDITING **162**
Cut editing
Use of audio tape recorders
SOUND AND ELECTRONIC EDITING **164**
Studio editing
Insert mode
Dub editing
DOLBY NOISE REDUCTION SYSTEM **166**
Dolby 'A' system
IN CONCLUSION **168**
The sound operation
The effect of the picture
FURTHER READING **169**

GLOSSARY **170**

Introduction

This book is concerned with the art and practice of television sound operations. It is intended primarily for the guidance of professional television sound operators and people involved in audio-visual operations and television productions of all types.

Sound reproduction, particularly when music is involved, is an art as well as a science, with considerable scope for individuality of approach. The presence of a picture provides an added dimension which can increase both the scope and the technical challenge of the sound operation. Because it is not an exact science it is not possible to lay down hard and fast rules for all sound operations. It is, however, important for all people concerned with the medium to have a clear idea of the scope available in each situation. This is the primary purpose of this book.

The treatment is essentially practical, technical information being introduced only where it is necessary for the proper understanding of the operation.

The method of handling each type of production situation is dealt with in detail, care being taken throughout to explain the underlying motivation and principles involved. The book should, therefore, be of interest to and understandable by production personnel, professional and amateur movie makers, users of audio-visial aids and anyone interested in the reproduction of sound in association with vision.

The book has six sections:

1. A general introduction to the sound medium with particular reference to the problems posed by association with the picture.
2. Types of microphone — the basic tool of the sound operator.
3. Methods of applying microphones for pick-up in various circumstannes.
4. Hardware on sound processing following the microphone and general matters of studio design and organisation.
5. Methods of tackling each different type of production situation, oroadly subdivided into a) the spoken word and b) music.
6. Peripheral equipment and processes through which the signal may pass after leaving the studio.

Acknowledgments

The author wishes to thank the Director of Engineering of the British Broadcasting Corporation for permission to publish this book.

Let us first consider the fundamental properties of sound.

The Nature of Sound

Sound is a vibration or rhythmic disturbance operating within the frequency range that we can hear. The lowest frequency that we can hear is about 15 hertz (cycles of repetition per second) and the highest about 20,000 Hz.

When it travels through air, which is the way we usually hear it, (we can also hear through bone conduction) it consists of waves of alternately increasing and decreasing air pressure, which travel outward from the source at a speed of around 340 metres a second. The actual velocity depends on the temperature of the air through which it passes, increasing with heat and, to a much lesser extent, humidity.

The velocity of sound is related to wavelength, i.e. the physical distance between one wavecrest and the next, in determining frequency (Velocity = frequency x wavelength). This is why orchestral wind players have to keep tuning their instruments, lengthening the air column as the concert hall warms up.

Sound intensity

The intensity of a sound depends on the power of the source and the manner in which it has travelled. If the source is very small, i.e. a point source, the wave front radiates as an expanding sphere and the intensity, (which depends on the area of the wavefront) diminishes as the square of the distance. If the source is large, or a long way away, the wavefront is practically flat (plane) and the intensity does not vary much with distance of travel. All this assumes that the sound is travelling freely in the open air, but in most cases it has to traverse rough surfaces where it loses energy by friction and absorbtion.

Loudness

When the sound reaches the listener his impression of its loudness depends not only on its intensity but also on many other factors, some of which are largely subjective and impossible to measure accurately. The factors include the environment of the listener and the ambient noise level, the pitch and character (harmonic structure) of the particular sound, the loudness of the sound that preceeds and follows it, and even such factors as the listener's appreciation of the programme material. Unwanted sound (noise) can sound louder than wanted sound of equal intensity. A discord can sound louder than a chord.

The subjective nature of sound

The more we learn the more we discover that sound engineering is not only a science. There is a strong element of the subjective, especially where broadcasting is concerned, which makes the task of the sound man at once interesting, challenging and creative.

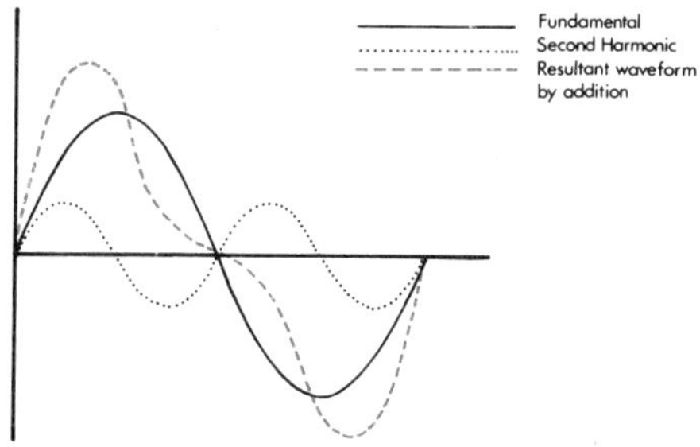

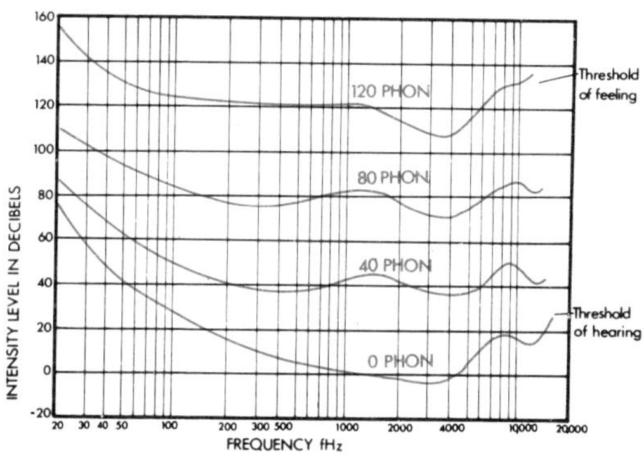

NATURE OF SOUND

Sound radiation
Sound radiates from a point source as a series of expanding spheres of alternating degrees of compression. At a distance, or if the source of sound is large, the wave front is practically flat.

Wave shape
The wave shape determines the character of the sound. This is what makes the difference, for instance, between a violin and a piano playing the same note. all sounds, no matter how complex, can be resolved into pure tones mixed in various proportions. These comprise the *fundamental,* the lowest frequency which gives the pitch of the sound, and the harmonics (multiples of the fundamental frequency) which give the sound its character.

Loudness curves
Curves of 'equal loudness' which illustrate the normal range of our hearing and the way in which the sensitivity of our ears varies with frequency and the volume at which we listen. Note how our hearing falls off at both ends of the scale, particularly at low volume. This is the reason for the inverse "psychology" compensation provided in many 'Hi- Fi' amplifiers. The scale is logarithmic because our hearing obeys a logarithmic law.

Nature of Television Sound

If you try the simple experiment of watching the television for an evening with the sound turned off three important facts will become obvious:

1. Notwithstanding the old adage "A picture is worth a thousand words", most of the information and substance of any television production is contained in the sound.
2. Most pictures tend to lose realism and impact when you cannot hear the accompanying sound.
3. You cannot concentrate, even on the most visual material, for any length of time unless you can also hear sounds that are at least in some way related to the picture. The old idea of the piano accompaniment to silent films was a crude example of sympathetic stimulation of the aural faculty.

Relative importance of picture and sound

Television is not just a visual medium; nor should it be said that the sound or the picture is the more important. In an effective production each is the perfect complement of the other. To say that either can stand up on its own is not necessarily to praise it. Rather does it suggest that the two media are not properly integrated.

It is all too easy to assume that the presence of the picture makes the achievement of good sound transmission unnecessary or less critical. This is not true. Most of the time all the picture does is to draw the attention of the viewer to what he is missing.

Relative complexity of picture and sound

It is also easy to assume that sound is the simpler medium. After all if you can see a source of sound, apparently at close range, you expect to hear it easily. With long-focus lenses, however, it is a simple matter to take close-up pictures from a considerable distance, but picking up the corresponding sound and separating it from the background noises could present an insuperable problem.

Comparison between sound and light

In many ways sound is the more awkward medium to control. It can curve around any obstacle of a size less than its wavelength, whereas light generally has to travel in straight lines. Sound also suffers many reflections, bouncing in all directions from any reflective object which is larger than its wavelength, such as the walls, floor and ceiling.

All this means that, while it is often possible to predict what a camera will see in a given direction, it is seldom possible to assess what a microphone will hear without taking into account many complex factors.

Control of light

Light travels in predictable straight lines. The picture from the camera is limited to the angle of view of the lens. The light which provides the source of reflected energy can often be controlled in intensity and confined in area.

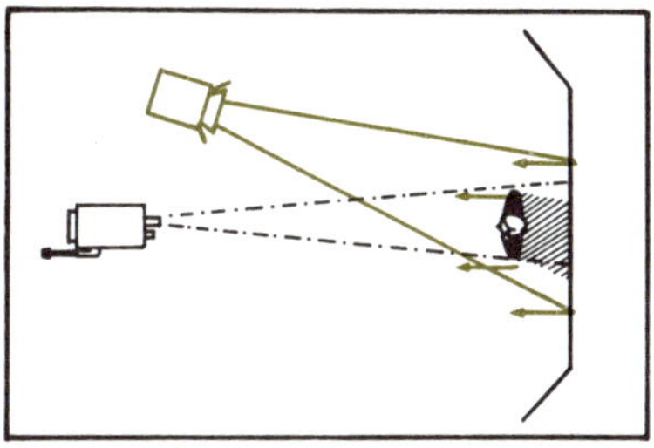

Control of sound

The sound source from the artist radiates over a wide angle and variable direction. The intensity is under the artist's control and the volume range often excessive and unpredictable. Sound reaches the microphone via direct and multiple reflective paths.

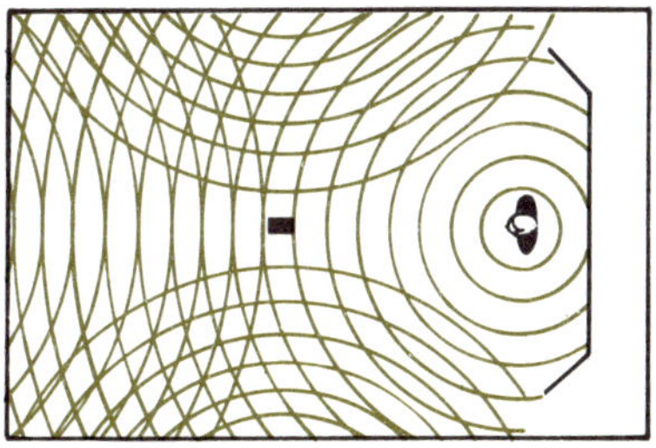

Unwanted sound

Noise or unwanted sound can reach the live face of the microphone via multiple reflective paths. Middle and low frequency sounds can curve around from behind sets or any other obstacle smaller than their wavelength.

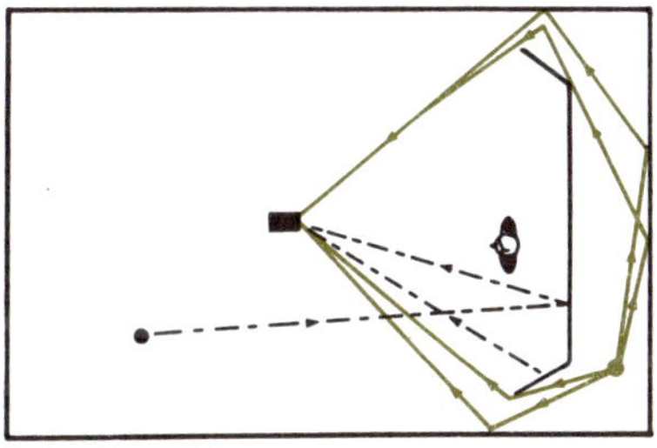

The sound supervisor is the aural equivalent of the vision director.

The Television Sound Supervisor

In television, the man in charge of sound is variously known as the Sound Director, Audio or Sound Supervisor. Throughout this book we refer to him as the sound supervisor.

He is commonly thought of as the person responsible for the *quality* of the sound, but this is rather like saying "a cameraman's job is to keep his lens clean". Certainly he is the custodian of technical quality, but he is rather more concerned with the *character* of the sound. His main task is to interpret the writer's or composer's intentions.

Reproducing sound for domestic listening is seldom a matter of producing a faithful copy of the original, because in most cases the listening conditions are so different. Some aspects have to be made larger than life and some smaller, with a delicacy of balance made acute by restricted listening levels.

The Sound Supervisor's first task is to suggest a suitable layout and design of the settings to produce acoustic enrichment for the sound and to promote the comfort and performance of the artists (see pages 70 and 74).

Sound balance and control

Having produced the right conditions for the production of the sound the next task is to balance it. The term 'sound balance' means the positioning of the artists and microphones and the mixing of their outputs to achieve a balanced artistic effect. It involves the exercise of choice on behalf of the listener and is the aural equivalent of camera direction.

In addition to achieving balance, the outputs of all sound sources have to be controlled for broadcasting or recording for three basic purposes:

1. To achieve an artistic balance between them from instant to instant.
2. To point emphasis — possibly even from a single source — by relating volume to importance.
3. To 'tailor' the original sound level variations to suit recording and broadcasting parameters and domestic listening conditions.

Staff control

The television Sound Supervisor commands a team assigned to each production that can include several boom operators, gram/tape operators and sound floor assistants.

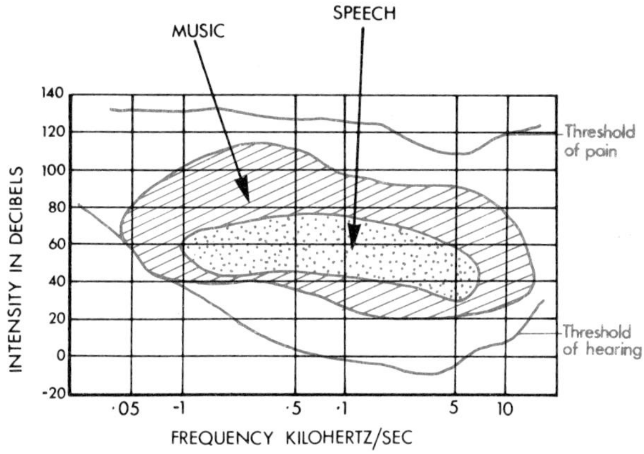

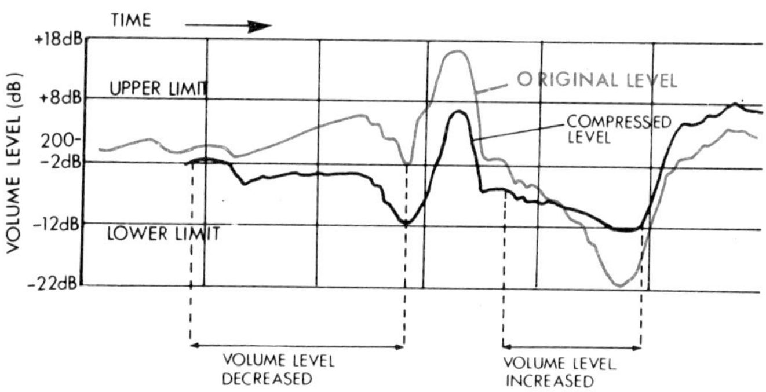

ROLE OF SOUND SUPERVISOR

Volume and frequency range

An approximation of the volume and frequency range encountered in music and speech, heard at normal distances, In practice with close-working multi-microphone technique, or in some drama situations, the actual level variations at individual microphones can be much greater. Some modern pop groups surpass the threshold of pain!

Anticipating peaks

The process of sound control requires a sensitive appreciation of the programme material and is achieved by intelligent anticipation with the aid of a script or score. Here, volume variations of about 40dB have been contained within a 20dB limit without losing impact. The operator has realised the importance of giving the sudden peaks their full artistic value and has made way for them by creeping the level down slowly when it will be least noticeable.

Pressure Operated Microphones

Most sound reproduction begins with a microphone of which there are various types. Pressure operated microphones consist essentially of a diaphragm enclosing a cavity that is sealed off from the surrounding air. They therefore respond to minute changes in air pressure and convert them into electrical signals.

A small leakage path is arranged (usually by means of a small bore hole between the inside and outside of the cavity) and this acts like the eustachian tube in the ear to allow the atmospheric pressure to equalise on a long-term basis.

Pressure operated microphones should have omnidirectional characteristics, i.e. they should pick up sound equally from all directions. In practice this is only true of the smaller and more expensive variety. The larger and less well designed examples usually have a response that becomes progressively more directional as the frequency is increased. This is due to the 'sound shadow' produced by their bodies.

Moving Coil Microphone

The moving coil microphone generates its output by means of a coil, attached to the diaphragm, which vibrates in a narrow annular gap between the pole pieces of a magnet. Movement of the coil within the magnetic field induces electrical current in the coil which provides the output. Some have coils wound to provide an impedance of about 30Ω, others have transformers built in to their cases to increase nominal impedance. Some moving coil microphones have an impedance that rises with frequency and require to work into about five times their impedance for a flat response.

Electrostatic (capacitor) microphone

In the electrostatic microphone the diaphragm forms one plate of a capacitor and the rigid back plate the other. The capacitor is polarised by applying a potential of about 60 volts through a high resistance. The vibration of the diaphragm results in corresponding changes in capacitance which are interpreted into voltage variations across the high resistance. These are applied through a blocking capacitor to a head amplifier close to the microphone element. In some examples various lengths of coaxial tube can be interposed between the element and the head amplifier to make them less conspicuous in shot.

Electret diaphragms

Some capacitor microphones employ electret diaphragms. This is a material (usually a high polymer plastic film) which has been given a permanent electrostatic charge. The need for a polarising voltage is thus eliminated and the assembly made much more compact.

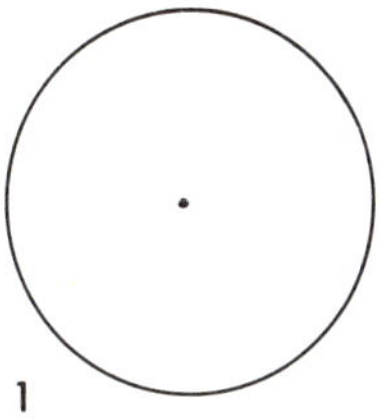

1

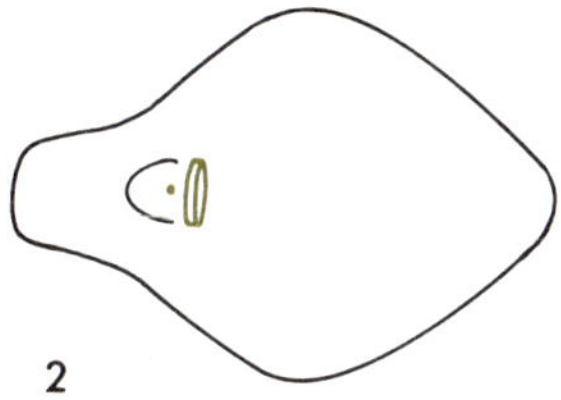

2

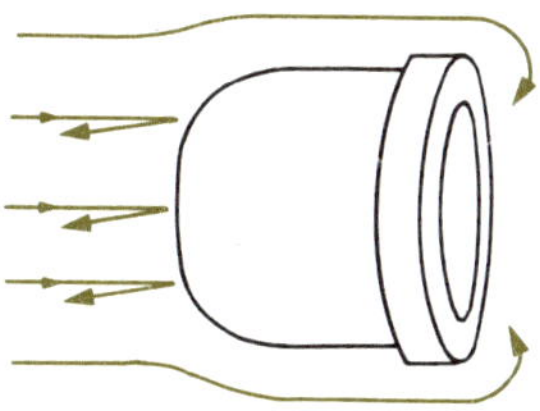

3

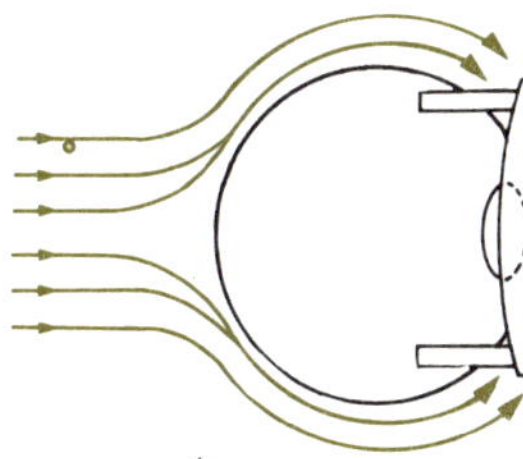

4

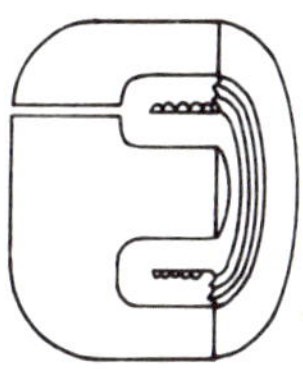

5

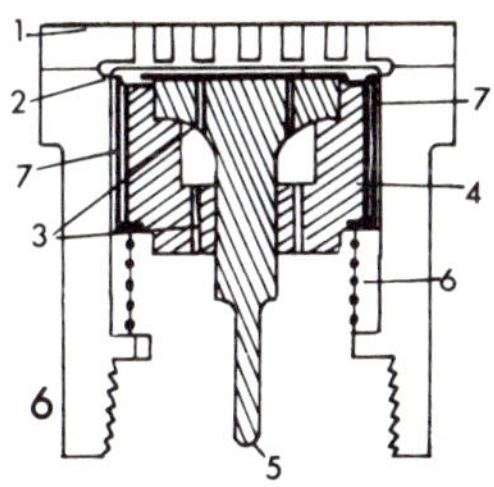

6

MICROPHONES

Pressure operated

1, Idealised polar diagram of presssure microphone. The distance of the line from the centre indicates the sensitivity in that direction. The diagram should be three-dimensional. 2, Typical H F polar response curve for large pressure microphone. 3, Sound approaching microphone from the rear showing how high frequency sound waves (wavelength less than dimensions of microphone) are reflected back and do not contribute to output. 4, Cylindrical microphone designed to reduce sound reflections by streamlining. Grille in front is to attenuate sound arriving from front to match with sound from other directions. 5, Half section through moving coil microphone showing diaphragm with corrugated edge with coil attached, and air vent to equalize pressure. 6, Section through electrostatic microphone showing the way the back-plate is slotted to allow air to escape as the diaphragm vibrates. *1*, Perforated front. *2*, Stretched diaphragm. *3*, Air holes through back plate. *4*, Insulator. *5*, Back plate contact. *6*, Spring connecting diaphragm contact to frame. *7*, Metal contact to edge of diaphragm.

Pressure Gradient Microphones

In the pressure gradient operated microphone both sides of the diaphragm are exposed to the sound waves. They make use of the difference in pressure that exists between the front and back of the diaphragm at any instant due to the extra time that the sound takes to reach the back.

Ribbon microphone

The most obvious example of the pressure gradient principle is the ribbon microphone. This consists essentially of a narrow ribbon of corrugated metal foil stretched between the pole pieces of a powerful magnet. As in the moving coil microphone (see page 16) movement of the conductor in the magnetic field induces electrical currents across the ribbon and these are applied via a step-up transformer (to increase the impedance) to provide the output. Sound waves have equal access to both front and back of the ribbon but waves approaching the microphone from the front obviously take longer to reach the back than the front of the ribbon. A difference of pressure thus exists at any instant in time owing to the difference in phase (time/wavelength relationship) between the two pressure waves (see page 10). This causes the ribbon to move away from the side of higher pressure.

Directional characteristics

Clearly the phase difference, and thereby the operating force, is greatest when the path-length difference is greatest, i.e. from front to back, and is non existent when the path length is the same, i.e. sideways on. This gives rise to the familiar figure-of-eight polar characteristics of the ribbon microphone.

Unidirectional Ribbon Microphones

Some ribbon microphones are made asymmetrical between front and back of the ribbon by partially restricting access to the back or causing the sound to reach the back via a longer path. The effect is to reduce the sensitivity at the back of the microphone and, by adding to the forward pressure component, to increase the sensitivity in the forward direction. In this manner ribbon microphones can be given a unidirectional (cardioid or hypercardioid) response, the details and advantages of which are discussed on page 20.

Unidirectional electrostatic microphones

Electrostatic microphones can also be made to have unidirectional response by allowing a proportion of the sound waves to reach the back of the diaphragm through an acoustic delay path (usually obtained by making the backplate of porous material or by drilling a solid plate with a complex pattern of holes).

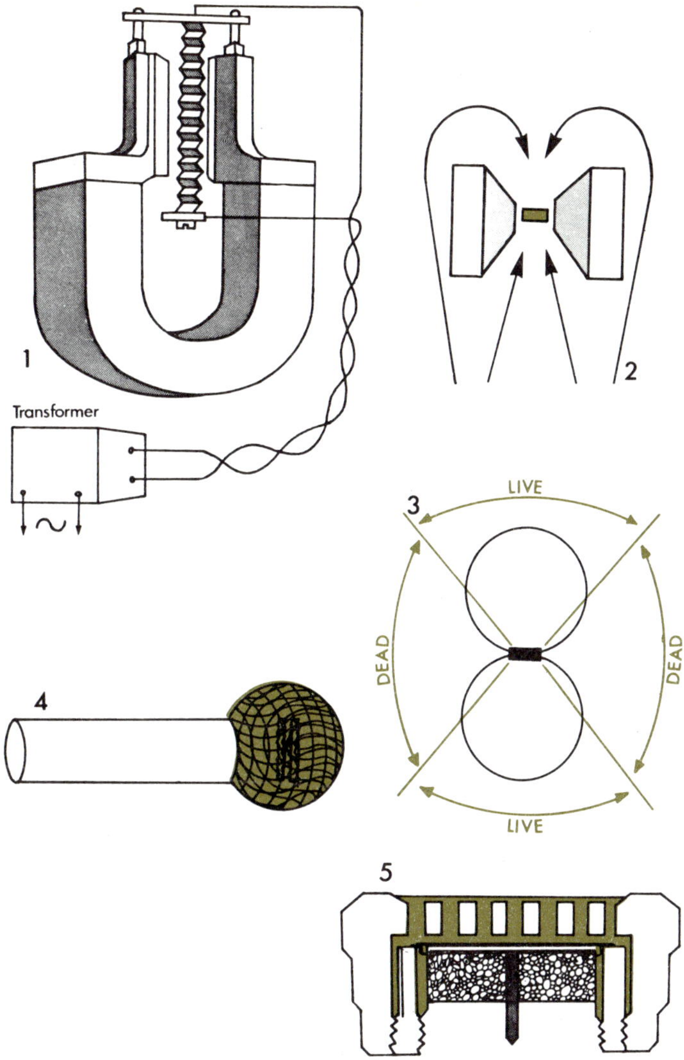

MICROPHONES

Pressure gradient
1, Outline of a ribbon microphone showing the corrugated ribbon stretched between the pole-pieces of a powerful magnet. 2, Plan view of ribbon microphone showing the extra distance that the sound has to travel to reach the back of the ribbon. The extra distance causes a time, and therefore a pressure, difference between front and back at any instant. This causes the ribbon to move. 3, The directional response of a ribbon microphone has a figure-of-eight polar diagram. Live front and back and dead at the side. 4, Cylindrical ribbon microphone. It employs a double ribbon at right angles to its length with an acoustic delay to the rear of the ribbon that give it a unidirectional response. 5, Unidirectional electrostatic microphone. The back plate is perforated and sound reaches it through a porous material which causes delay.

Double Element Microphones

Microphones with variable directivity characteristics are produced by varying the proportions of pressure and pressure-gradient operation, either by changing their physical characteristics or, more usually, by employing more than one element and varying the proportion of each output.

Combination of pressure and pressure-gradient types

If we imagine the output of a pressure microphone (omni-directional, circular polar diagram) added to an equal output from a pressure gradient microphone (bi-directional, figure of eight polar diagram) it will be seen that the outputs add (in phase) in one direction and subtract (out of phase) in the opposite direction. Plotting the addition of the two characteristics produces the familiar cardioid characteristic.

If a graph is plotted of the directional combination of pressure and pressure-gradient elements while the proportion of each is varied, two very important facts emerge, both of which are of vital significance to the users of microphones:

1. From the point of view of discrimination between incident and ambient sound, e.g. noise or reverberation etc. (assuming it comes equally from all directions), a cardioid and bi-directional response are equal with a discrimination factor of 3:1 (curve A).
2. The most directional condition and best discrimination between direct and ambient sound is achieved in the hypercardioid condition i.e. a 'cottage-loaf' characteristic with a narrower than cardioid front response and a small lobe at the back. This gives an overall discrimination ratio of approximately 4:1.

Back-to-back electrostatics

One type of microphone in common use has directional characteristics that can be varied remotely between omni-directional through cardioid to figure-of-eight. This is achieved by mounting two electrostatic microphones back-to-back (in fact sharing a common back plate and varying the polarising potential and polarity of one of the plates with respect to the other).

Another type of double-element microphone has been produced which employs two unidirectional elements, one for the treble, and one for the bass with a cross-over frequency of about 500 Hz. This principle helps the designer to overcome the acute problems of obtaining the optimum phase relationship between the front and back of the microphone throughout the frequency range and thereby obtain a smooth and consistent frequency response and directional characteristic.

1
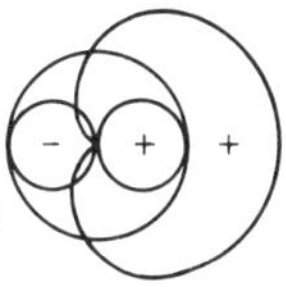

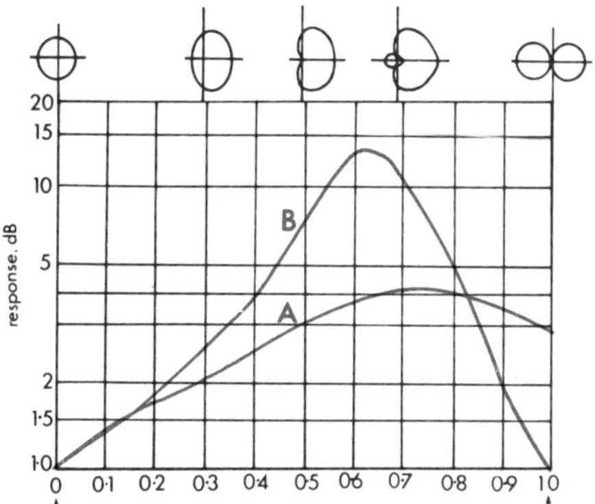

2

3
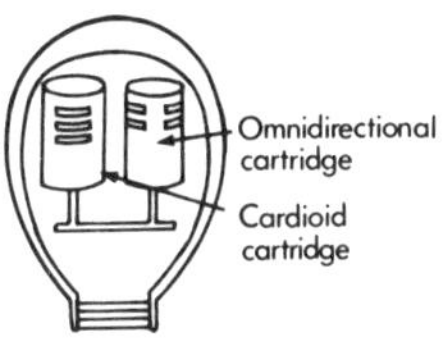

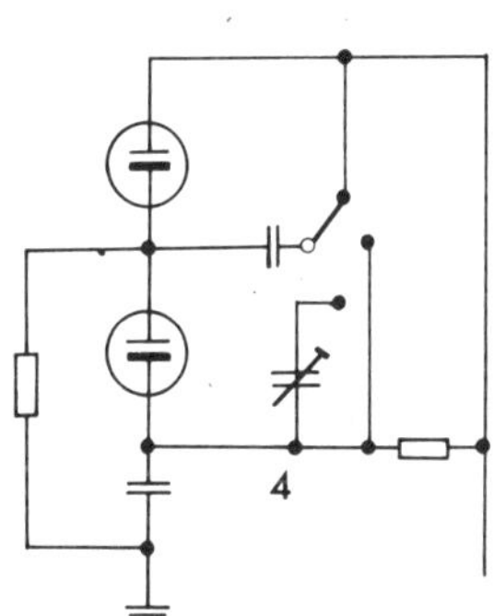

4

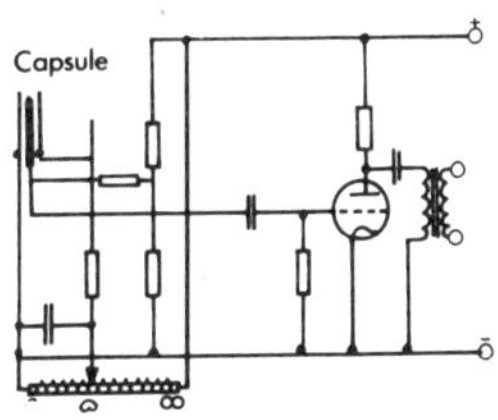

5

6

MICROPHONES

Double element

1, If a circular polar diagrams is added to a figure-of-eight the result is a cardioid (heart-shaped). 2, Limacon curves, showing the effect of adding different proportions of omnidirectional and bi-directional responses. Curve A gives the result in terms of directivity (i.e. ratio of incident to ambient response). Curve B shows the unidirectional factor (i.e. ratio of response in forward direction to all other directions). 3, Dual element electrostatic microphone showing separate cardioid and omnidirectional elements. 4, Method of connecting dual element microphone to give three different responses. 5, Method of connecting double-element microphones with unidirectional elements back-to-back to provide variable directivity. 6, Double element moving coil microphone.

Microphones that have to be kept out of the picture need super-directional characteristics.

Super-Directional Microphones

The best overall discrimination between incident and ambient sound from all directions that can be obtained from normal microphone design is 4:1 (the hypercardioid "cottage-loaf" pattern).

Greater selectivity, to enable the microphone to be used at a greater distance or against a loud random background noise, requires some sort of focusing device.

The reflector microphone

The simplest and most efficient method of producing a very narrow angle of acceptance is to place the microphone at the focus of a parabolic reflector. This arrangement is very efficient because the reflector acts like a large scoop which collects the sound and focuses a narrow beam on to the face of the microphone. The best type of microphone to use for this purpose is one with cardioid response arranged so that the live face looks into the reflector, encompassing all of it in its acceptance angle, and the back faces the wanted source of sound. This may seem odd until it is realised that it is a feature of the parabolic shape that the path lengths of all the waves from a source straight ahead to the point of focus are equal, so that they reach the microphone face in phase. Any sound picked up direct would travel a lesser distance and, being out of phase, would tend to cancel rather than augment the reflected sound.

Reflectors can only be really effective for sounds whose wavelength is less than their diameter.

Even a one metre dish, which can only be effective for frequencies above 300Hz, can be a very cumbersome piece of equipment, especially out of doors in a high wind.

The rifle microphone

An alternative method of producing a narrow angle microphone is to couple it to a length of tube with a slit or series of holes running the complete length. When the tube is pointed at a source of sound the waves travel straight down the tube and actuate the microphone. Sources of sound reaching the tube from other directions not only go straight down the tube but enter through each of the holes or at different positions in the slit. They thus arrive at the microphone diaphragm in differing phase, having travelled different distances, and tend to cancel each other out. Hence the term "interference tube". This type of microphone achieves its narrow angle by rejecting unwanted sounds rather than by efficient collection of wanted ones as in the case of the parabola, so that it tends to lack sensitivity when required to be used at great distances. Modern examples of interference tube rifle microphones usually employ electrostatic microphones with integral head amplifiers to mitigate the problem.

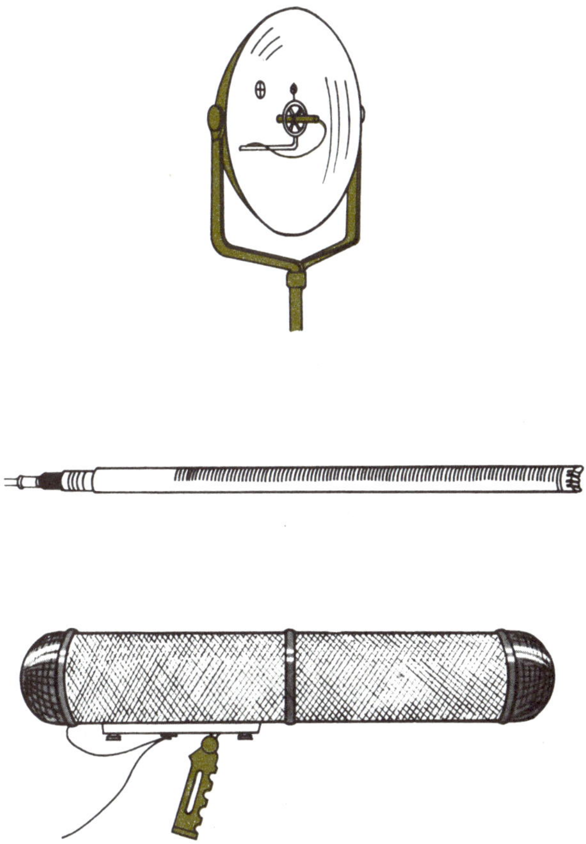

MICROPHONES

Super directional
Top: Microphone in parabolic reflector. The unidirectional microphone should be facing into the dish and away from the source of sound. Note sight hole and cursor for aiming the dish. *Middle:* Rifle (interference tube) microphone. Note sound entry slits down one side. *Bottom:* Rifle microphone with windshield and pistol grip.

Stand and Hand-Held Microphones

The use of a stand microphone in television suggests that:

1. The artist is static.
2. The microphone will come into shot.

A very important feature of a stand microphone, therefore, is its appearance. The microphone should be slim and neat and preferably mid tone in colour with a matt finish to minimise specular reflections from the lights. For the majority of circumstances a unidirectional response is required (either cardioid or hyper cardioid) so that sufficient separation can be obtained from unwanted sound or accompaniment without the necessity to work so close that the microphone masks the artist or unduly restricts his movement. Experience suggests that 'end-fire' microphones are better for this purpose than side facing ones because artists find it reassuring to have the microphone pointing toward them. (A side-facing microphone in a low position has to be angled away from the artist).

Microphone stands are normally telescopic and can be made or cut down to a variety of sizes. Normally, three sizes are useful for television: 1ft to 1ft 6in (0.3m to 0.45m) for low-speaking instruments such as the double bass and guitar amplifiers or for inconspicuous operation; 2ft to 3ft 6in (0.6m to 1.0m) for seated artists; 3ft to 5 ft (0.9m to 1.5m) for standing artists. These larger versions can also be used in association with small 'lazy arms' which are particularly useful for reaching across, for example, pianos or music stands. Extreme neatness and unobtrusiveness can be provided by the extension tubes available for electrostatic microphones (see page 16). These are thin co-axial tubes that can be interposed between the microphone capsule and the head amplifier, thereby presenting a very slim and unobtrusive appearance for the part of the microphone that is likely to come into shot. The tubes are curved at the top to provide an angle for the capsule which cannot be adjusted unless a special swivel joint is used between tube and capsule.

Hand held microphones

When the appearance of a microphone in shot is not objectionable but the use of a stand would restrict the movement of the artist a hand-held microphone can be used.

The requirements for a hand-held microphone are that it should be slim, long enough to hold conveniently without the hand covering any sound entry ports and as immune as possible from handling noise.

It should not suffer unduly from low-frequency rumble or wind noise when moved through the air and the cable must be sufficiently flexible so as not to cause rustles. If the microphone is to be used very close to the mouth, as is usually the case for pop vocalists, it is usually advisable to choose one with omnidirectional characteristics, because it is less likely to produce unpredictable volume variations.

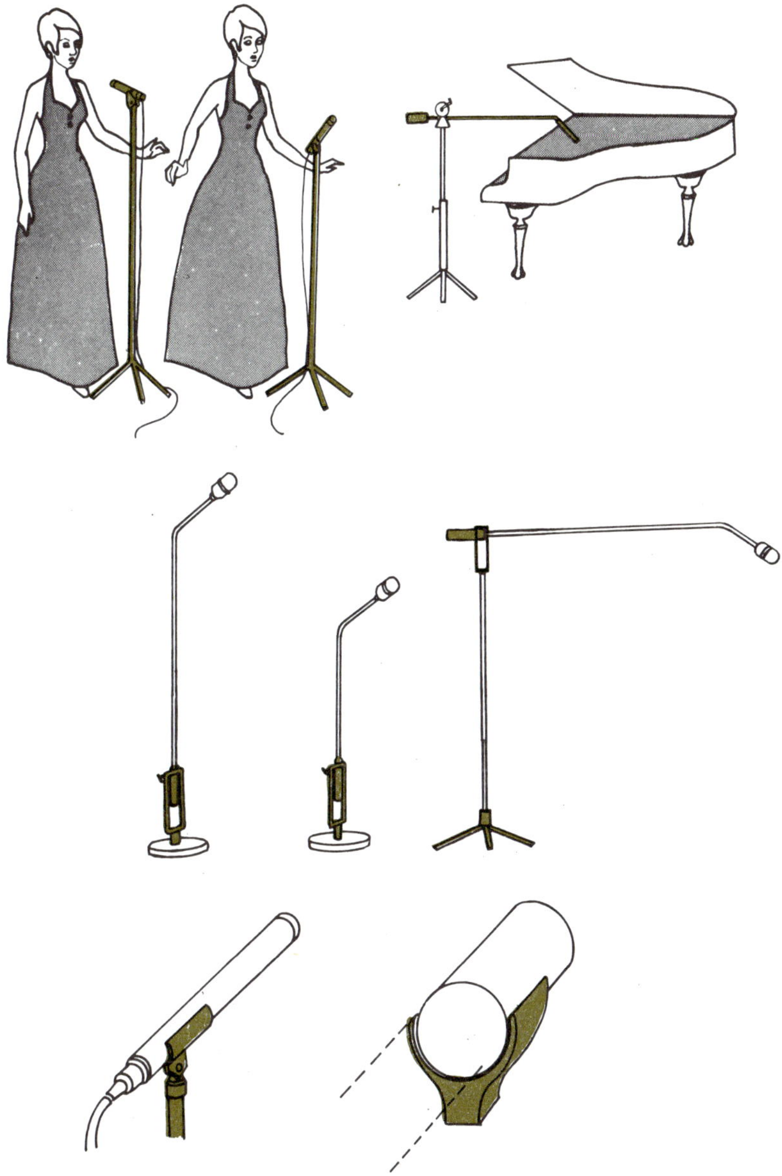

MICROPHONES

Stand and hand-held

In general 'end-fire' microphones *(top left)* look better on stands than those with side sensitivity which, when used low enough to be unobtrusive, lean away from the artist. Small lazy arm *(top right)* is especially useful for musical instruments. *(Middle)* : Electrostatic microphones with extension tubes. Microphone clip *(bottom)* allows the microphone top be removed from the stand and hand held.

Lavalier Microphones

One method of picking up sound from an artist with a limited range of movement is to use a personal microphone either suspended on a lanyard around his neck or clipped into the clothing.

Special microphones are designed for this purpose (sometimes called lavalier or lanyard microphones). The requirements for a lavalier microphone are small size and particularly slimness and lightness in weight. The connecting cable must be light and very flexible so as to minimise the production of rustling noises as it moves against the clothing. Such a cable will not be sufficiently robust to be allowed to trail across the floor and it is normal to terminate the thin cable at about one metre with a suitable connector so that it can be joined to a normal floor cable. This provides the added advantage that the cable can be threaded through the artist's clothing easily and cleanly, the junction being taped to the waist, or to the leg if it is threaded through trousers.

Supplying necessary HF lift

Some lavalier microphones are designed to have special amplitude/frequency characteristics to compensate for the way they are used. The usual position that a microphone adopts when slung round the neck is a very unsuitable place from which to pick up speech. It is screened from the mouth (which projects the sound away from it) by the chin but in line with the projection of nasal tones. It can become obscured by the clothing (which reduces high frequencies). For these reasons some lavalier microphones are given rising amplitude/frequency characteristics. A type of built-in 'presence filter' (see page 66).

Unfortunately the degree of HF lift required varies with individuals and circumstances so that it may be more satisfactory to employ a 'straight' microphone, e.g. the electrostatic type, which can have an impeccable frequency response, and adjust the output by means of a variable filter if necessary. One type of dynamic lavalier microphone offers the facility to vary its frequency response by adjusting the projection of its clip. When the clip extends over the end of the microphone it produces a small cavity resonance which provides a HF boost peaking in the 6 kHz region.

Miniature electrostatic microphones

Electrostatic microphones are available with miniature heads and diaphragms only a few mm in diameter, in some cases using the electret principle (see page 16). This is followed by a compact head amplifier at a convenient distance. Some provide excellent quality and are eminently suitable for lavalier use.

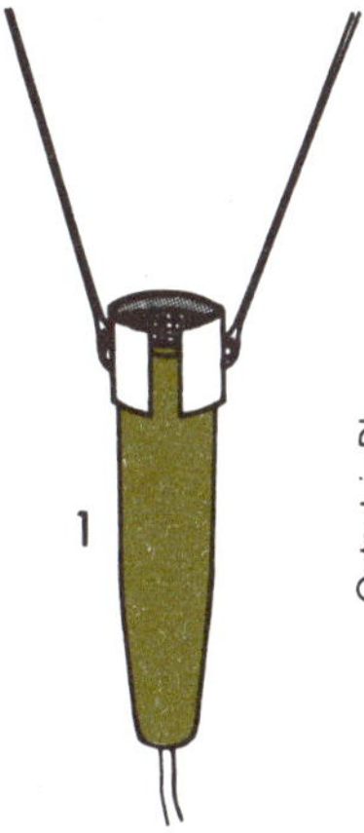

1

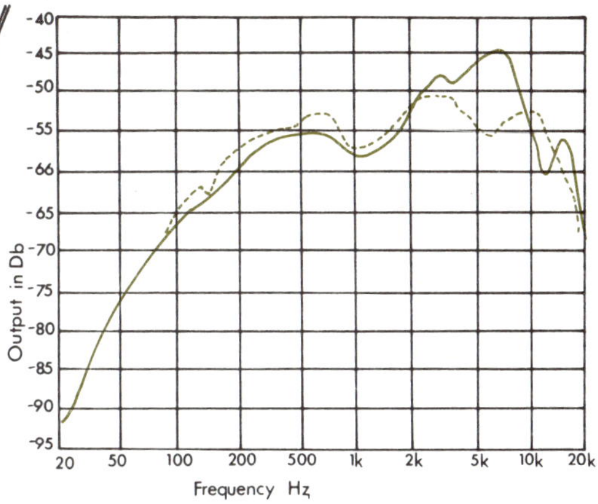

2

3

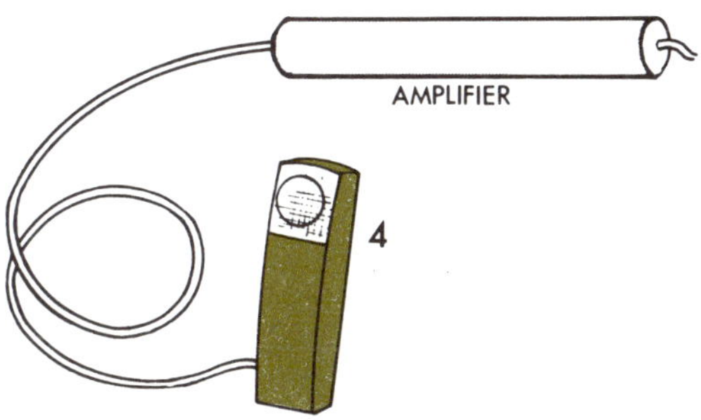

4

MICROPHONES

Personal or lanyard

A personal lanyard microphone incorporating a movable clip. 2, Illustrating the effect of moving the clip. When the clip extends over the top of the microphone a cavity resonance effect causes an increase in the response at around 6KHz giving increased 'presence'. 3, A miniature electrostatic personal microphone which uses an electret diaphragm. 4, An electrostatic micrtphone in which the diaphragm faces away from the artist wearing it. The idea is to give preference to the other person in an interview situation.

Radio Microphones

When an artist has a large area of movement, or when a cable would be an encumbrance a radio microphone can be used. This usually consists of a small transmitter, carried on the person of the artist, connected to a personal microphone (either lavalier or hand-held). The transmitter is usually carried in the pocket or in a pouch tied around the waist. There are some radio microphones in which microphone and transmitter are built into the same case in the form of a baton.

Transmission systems

The frequency-modulated signal is picked up by a receiver which is normally housed in the sound control room. The choice of receiving aerial depends on the circumstances. If only one artist, with a limited range of movement, is involved, a single dipole aerial can be used, mounted as near as practicable to the action. If several artists with radio microphones are required to move about extensively, it may be advisable to work their respective aerials at some distance (e.g. 10 metres away from the action) and well separated from each other to avoid 'frequency pulling' of the receivers by the other transmitters.

Long range reception (e.g. an operator walking with a gun microphone and transmitter to cover sound effects for a golf match) may require a highly directional aerial panned to follow the action. In difficult circumstances, where large areas of movement are involved and there are multiple reflections from moving cameras and other metal structures, a form of 'diversity reception' is required to avoid the effect of 'nodes' (positions of signal cancellation) due to out-of-phase reflections. This can be a simple matter of combining a number of aerials into one receiver, or a system of switching between two aerials at a supersonic rate, the theory being that the signal should be available from one or other of the aerials at any given instant.

The receiver should have a limited A.G.C. or a muting circuit to prevent sudden increases in noise level in the absence of signal, which is much worse than momentary silence. In the ultimate a sophisticated diversity receiver would be used which measures the input of several aerials and automatically switches to the best one from instant to instant.

Site surveys

Whenever possible the area in which radio microphones are to be used should be carefully surveyed in advance for signal strength and possibility of interference. When on outside broadcasts, watch out for large transmitters in the vicinity, police and other radio intercom systems, neon lighting and other sources of electrical interference.

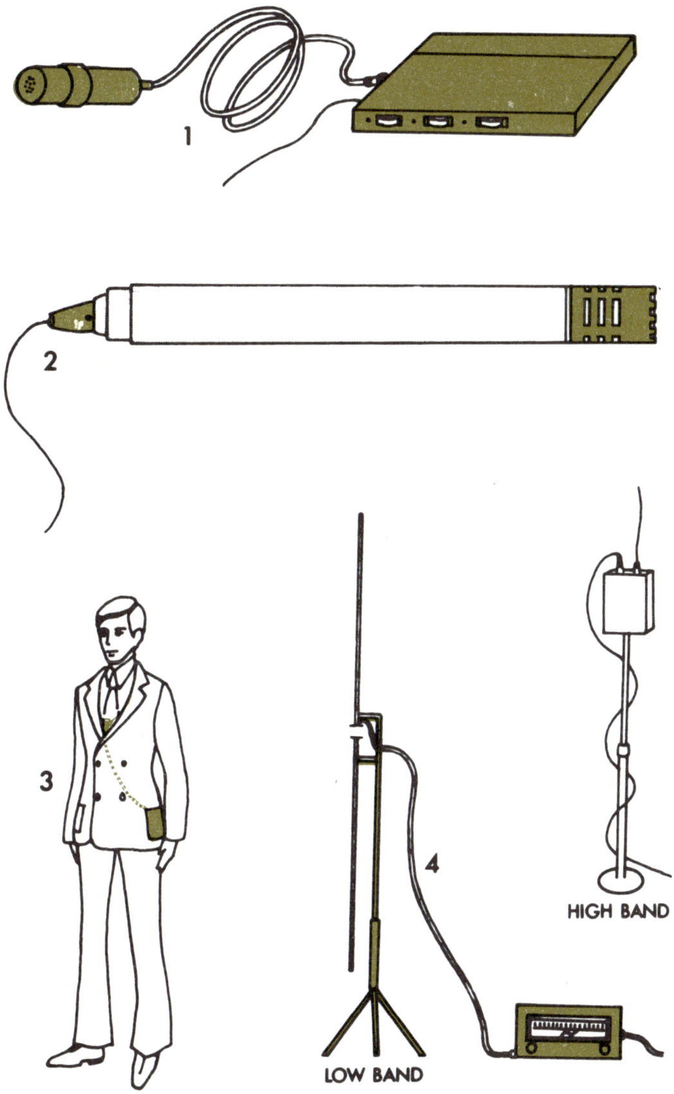

MICROPHONES

Radio-operated
1, Radio microphone for studio use. The transmitter is miniaturised and made as inconspicuous as possible. The microphone can be any of the types described on the previous page. 2, Baton microphone for interviews. This type can be handed from person to person easily. 3, Method of accommodating a radio microphone. 4, The aerial should be as close as possible to the action. A portable stand can be useful.

Microphones mounted on short poles can be useful for covering awkward areas.

Fishing Rods

In circumstances where the action covers a limited area and microphones are not to be seen in shot but the use of suspended or hidden microphones (see page 34) would be awkward or too inflexible, fishing rods can be useful. These can consist of light bamboo or aluminium poles, about two metres long, with a lightweight microphone mounted on one end.

Type of microphone

The best type of microphone for this purpose is usually an electrostatic cardioid, which can combine light weight with good directional characteristics. However, if the fishing rod microphone is used only for a short sequence to take over from some other method such as a boom, it might be necessary to employ the same microphone type as used on the boom to preserve continuity. Nowadays electrostatic microphones are often used on booms so there should be no problem in matching the types.

Microphone suspension

The microphone should be suspended in a suitable mounting and if necessary provided with some windshielding. The degree of windshielding required depends on whether the action is out of doors with a possibility of real wind or if it is necessary merely to guard against wind turbulence due to the microphone being moved rapidly through the air. In this latter case a relatively small close-talking shield is adequate. Protection against strong wind requires a much larger windshield (possibly in excess of 10 cms in diameter) to reduce the effect of minor turbulences over its surface reaching the diaphragm. It is important that the microphone is rigidly clasped inside the windshield and the shock suspension applied overall otherwise movement inside the shield as the rod is swung about would cause its own air turbulence.

Operation

Operator fatigue can be reduced by providing a counterweight at the other end of the rod to form a balance at the point of holding. If the sequences are longer than a few minutes and the rod has to be held above the action a shoulder bracket can be an asset. One of the main advantages of the Fishing Rod particularly when used in confined spaces is the ability to use it below, or to the side of the shot.

Line (rifle) microphones equipped with pistol grips can be excellent for fishing rod operation especially when the optimum approach is from the side or under the shot. The use of a radio transmitter with a fishing rod or hand-held line microphone can greatly increase operator mobility.

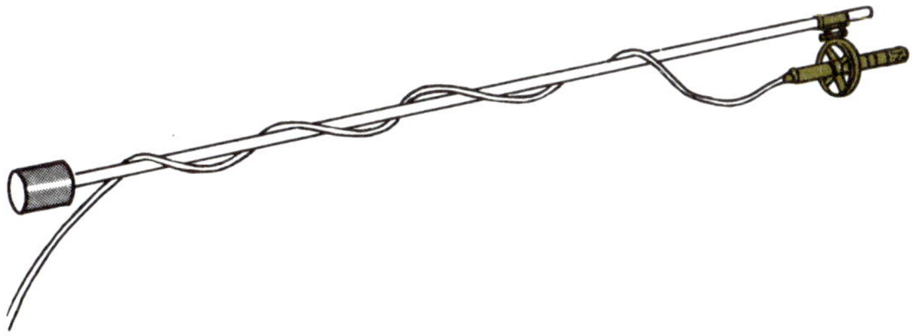

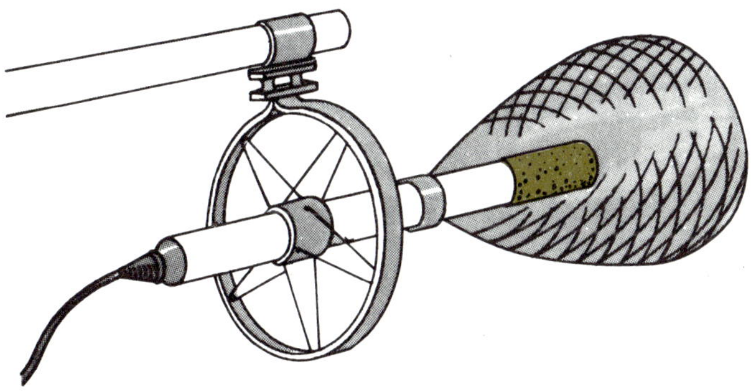

MICROPHONE SUPPORTS

Fishing rods

Top: Fishing rod, with counterweight at back end. *Bottom:* Detail of suspension mounting. The windshield must be large for outside use and fixed to the microphone, not the suspension.

Some areas can be covered by microphones slung overhead.

Slung Microphones

When it is required to cover a source of sound by means of a static microphone that does not appear in shot, it is generally best to place a slung microphone over the action. By the nature of things the source of sound, e.g. the artist's head, is usually closer to the top of the frame than any other edge.

The choice of a slung microphone suggests that the action is static. For fluid action several slung microphones are used, continually fading from one to another to cover movement or even changes in the angle of the artist's projection. This is not usually a satisfactory arrangement, because each microphone can produce a rigging problem and the risk of getting itself, or its shadow, in shot, as the action progresses.

Microphone in shot

Slung microphones can look very ugly in shot and the fact that the point of suspension is above the frame of the picture can leave an uncomfortable query in the mind. They can be acceptable in long-shot providing the suspension is kept as neat as possible. Often the neatest method is to employ electrostatic microphones (see page 16) with the one metre extension tubes hanging upside-down. If the lower end of these appear in shot at some distance it is likely to be very unobtrusive especially if it is covered with a dark stocking to avoid catching specular reflections from the lights.

An even neater arrangement is to hang an electrostatic personal mic upside-down by its cable, provided that directional qualities are not required.

Method of suspension

It is usually advisable to sling microphones from three points of suspension so that a wide range of adjustments can be made in the light of rehearsal experience.

Slings, or at least the last few metres at the microphone end, should be made of thin but strong material such as nylon or steel wire. Whenever possible elastic suspenders should be interposed between the sling and the microphone to prevent the transmission of sound along the sling reaching the microphone.

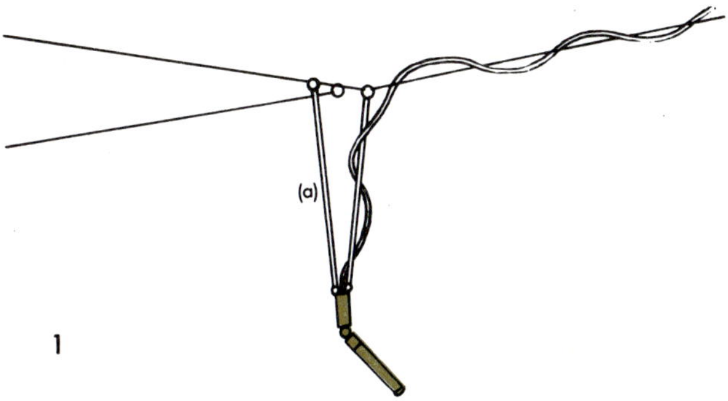

1

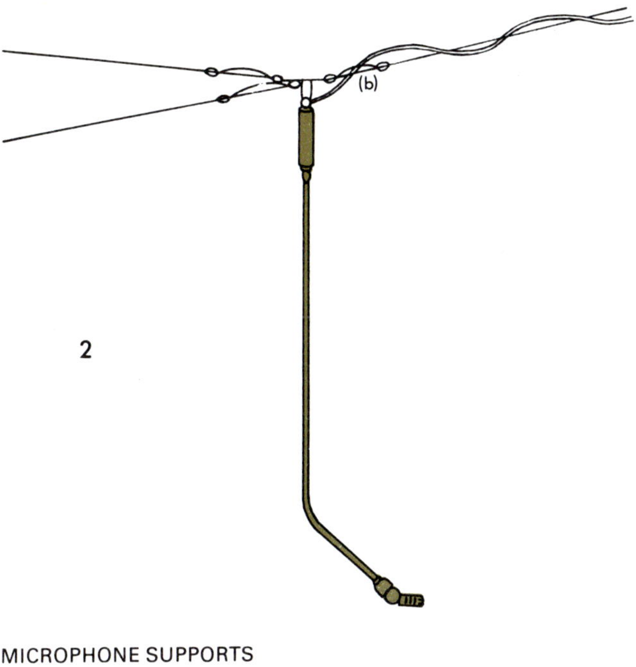

2

MICROPHONE SUPPORTS

Slung microphones
1, Whenever possible microphones should be slung from a three-point suspension so that they can be adjusted in all directions. 2, An electrostatic microphone with extension tube slung upside-down makes a very neat suspension which is hardly visible in a long-shot.

Suspension methods
It is always advisable to incorporate elastic suspension to prevent vibrations, transmitted down the slings, reaching the microphone. In the slinging method shown in 1(a) the rubber suspenders form the vertical elements and merely take the weight of the microphone. In the method shown in 2 (b) the suspenders are incorporated in the slings and have to take the full tensional strain. For safety reasons, therefore, they should be by-passed by slack nylon cords.

Microphones can be hidden in objects that are acoustically transparent.

Hidden Microphones

If it is desired to cover some part of the action by an out-of-shot fixed microphone but to sling it would result in its being too far away it is always worth considering the possibility of using a hidden microphone.

Maintaining microphone characteristics

The most important thing to remember about hiding microphones is that the material in which they are hidden must be acoustically transparent i.e. sound must be able to pass right through it with a minimum of reflection or absorption. Otherwise, the response of the microphone is modified and the sound is so unnatural that the stratagem is exposed.

Suitable material for surrounding a microphone is fine mesh metal or plastic gauze or fairly open weave cloth. If this is light in colour and shaped to represent some solid article such as a vase or pile of books etc. the microphone can be hidden very effectively and its operation will be unimpaired. It is also possible to hide the microphone within a flower arrangement or in a lampshade etc. If there is a problem in hiding the cable, a radio microphone (see page 28) can be used, but care must be taken not to exhaust the battery by forgetting to turn it off between takes or rehearsals.

An ideal type of microphone for hiding would be the electrostatic personal type (see page 26), which can be very small and unobtrusive. They are omnidirectional so do not need to be angled carefully, but must be used at fairly close range to discriminate against noise.

Sound perspective

Whatever type of microphone and hiding place is chosen, great care must be exercised to ensure that it comes roughly on a line between artist and camera and that the artist projects his voice in that direction to avoid reverse perspective effects the artist appearing to recede in sound as he approaches in vision and vice versa.

1

2

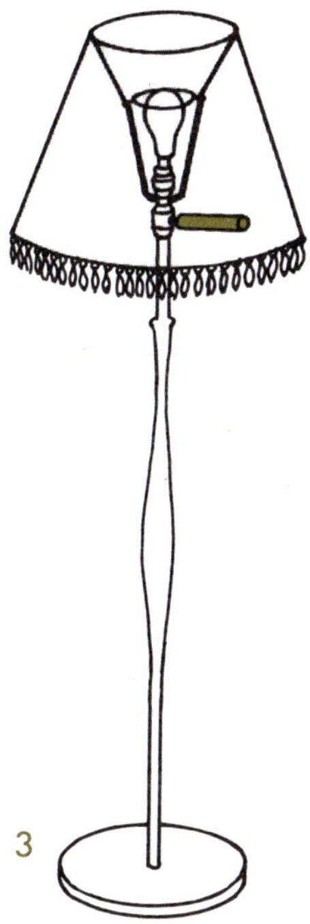

3

MICROPHONE SUPPORTS

Hidden microphones

1, Vases of flowers can be useful for hiding microphones on a table, possibly between two people in dialogue. The two unidirectional microphones illustrated would be used so as to balance between the voices. 2, A table lamp can be constructed out of fine wire mesh covered with gauze. If it is light in colour it looks solid and the microphone is invisible. 3, Standard lamps can also provide useful hiding places provided the shades are of acoustically transparent material.

Sound Booms

There are several types of microphone booms but essentially they consist of a long telescopic pole with a microphone on one end and a counterweight on the other. This is pivoted on a vertical column on 'the pram' which is mounted on wheels and usually provides a raised platform and in some cases a seat for the operator to see the action above the normal height of the cameras.

Operation of the boom

The length of the arm can be adjusted by means of a 'racking handle' which operates through a series of pulleys. The weight of the microphone on one end of the boom arm is counterbalanced by a much heavier lead weight at a lesser distance from the fulcrum on the other end. As the telescopic arm is racked out the counterweight moves back a corresponding amount so that the balance is uniformly maintained. In addition to racking the microphone in and out the operator can also swing the arm through an arc and swivel the microphone by remote control in both the horizontal and vertical plane. This is achieved by a handle, held in the left hand, which is raised or lowered to twist the microphone sideways and squeezed to tilt it up and down. The handle is mounted to the rear of the pivot point and the handle operating the rack mechanism, which is operated by the right hand is in line with the pivot so that all four actions can be accomplished simultaneously.

The boom pram

The boom pram is normally fitted with three wheels and some are capable of being crabbed (i.e. all wheels steered parallel to each other) or tracked (the rear wheels only being steered). Some types of boom provide the ability to alter the height of the arm fulcrum and the platform (the relationship between these two must be fixed) either by a winding mechanism or by hydraulic operation. This latter arrangement enables the adjustment to take place even during the action, making it possible to dip down to track under a low door or to rise above a camera shot.

Typical boom dimensions are as follows:

Pram width 4ft (1.2m), Length 5ft (1.5m), Boom arm maximum extension 20ft (6m), Tail length 4ft (1.2m), Minimum extension 10ft (3m), Maximum microphone height 15ft (4.5m), Minimum overall height 8ft (2.5m).

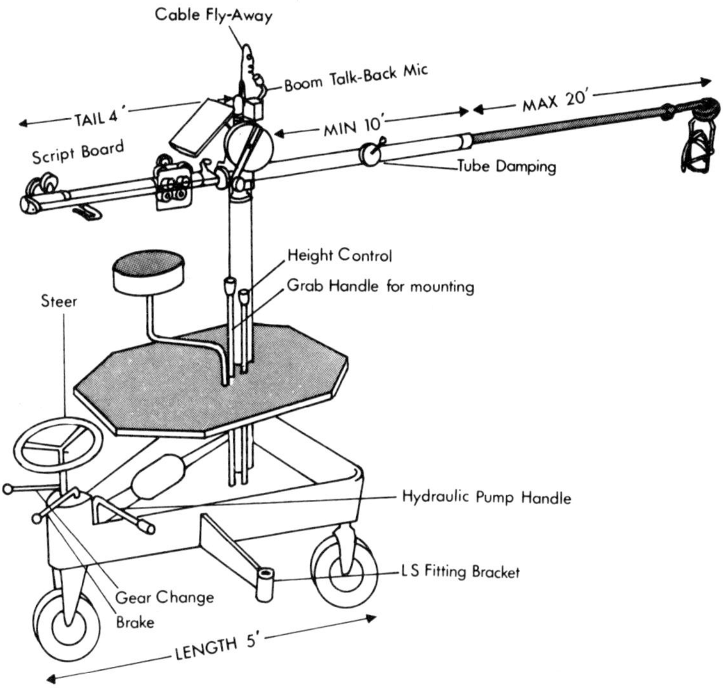

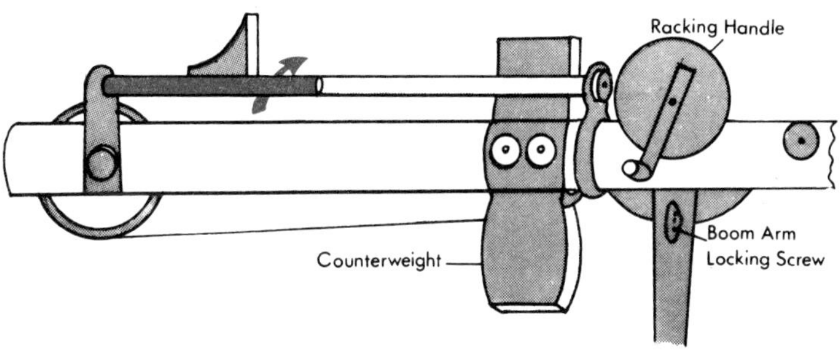

MICROPHONE SUPPORTS

Microphone booms
A boom suitable for most studio applications. The height of the platform and arm is adjusted by an hydraulic ram controlled by a lever on the platform. The revolving seat is intended for occasional use. It is difficult to do rapid swings and complicated manoeuvres without standing up. *Bottom:* Detail of back end of boom arm showing operational control.

Boom Microphones

The most important point to remember when selecting a microphone for use on a boom is that for most of the time the microphone will be operating at a distance greater than the optimum.

Directional characteristics

The directional characteristics of the microphone are of special importance as there is a continual requirement to discriminate between wanted and unwanted sound such as noise, orchestral accompaniment (which should be picked up only by microphones properly positioned for the purpose), and reverberation etc. This discrimination between wanted and unwanted sound must be good enough to allow the microphone to work at a sufficient distance so as not to cramp the camera headroom.

Moreover, as it is inevitable that a proportion of the accompaniment will be picked up on the dead side of the microphone it is important that it has an even off-axis response, i.e. that the 'dead' areas are equally 'dead' over the full frequency spectrum. This is especially true in the case of music where, unless the microphone can be used sufficiently close to the artist to achieve discrimination by exploiting the relative distance of the sources, a proportion of the accompaniment will be picked up on the dead side of the microphone. In these circumstances it is usually worthwhile to employ a good quality electrostatic microphone for the boom provided that a suitable shock mounting and sufficient wind shielding can be provided.

Acceptance angle

The most useful acceptance angle for a boom microphone is likely to be cardioid. This gives a reasonable compromise between the narrow response necessary to give sufficient discrimination at a reasonable working distance from the artist and the ability to pick up dialogue from several artists spaced apart by several metres without making the aim too critical.

Although the microphone can be tilted and turned in all directions there is seldom time between speeches to re-aim it accurately. Where exceptional working distances are involved, or the need for separation is great, e.g. in audience participation programmes, (see page 90) rifle microphones (see page 22) can be used.

Construction and mounting

Microphones used on booms must be robust to stand up to continual shaking. Even so they must be provided with effective shock mountings. Effective windshields are also necessary if boom microphones are to be moved through the air rapidly and especially if they are to be used out of doors.

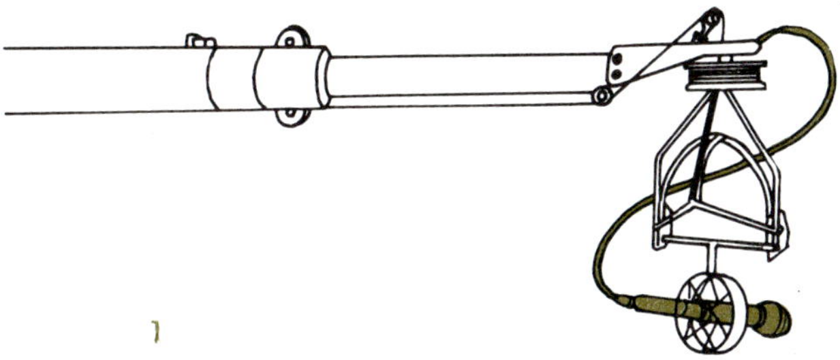

1

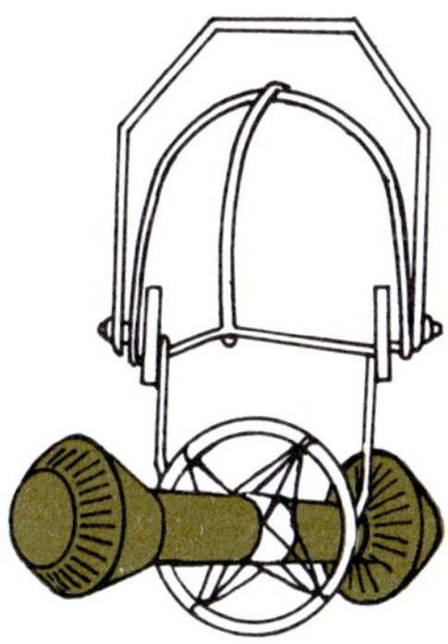

2

MICROPHONE SUPPORTS

Boom microphones

1, Detail of end of boom arm showing method of adjusting 'pan' and tilt of microphone. The microphone is mounted in rubber suspenders within the tilting cage which is also mounted on soft suspension. 2, Double element microphone in use in boom showing the need to fit windshields at both ends.

The boom can cover a large area of movement provided certain conditions are met.

The Scope of the Boom

In every production situation there is almost certainly one optimum position for each microphone. If a microphone is to be used to follow free-flowing movement, it too should be mobile and the best way to achieve this is with a microphone boom.

Advantages
The range of mobility varies with the type of boom but a modern full size boom allows the microphone to be racked in and out, swung from side to side and tilted up and down, thereby covering sound within a segment of an arc of almost 180° in the horizontal plane and almost 30° in the vertical plane with a radius between about 4 and 7 m (assuming the artist is about 1m in front of the microphone).

The microphone boom can also be tracked or crabbed, i.e. moved sideways with its wheels in parallel track, so that its range of coverage is limited only by practical considerations. The microphone can be turned through about 300° horizontally and about 90° in the vertical plane and all these movements can be accomplished simultaneously so that changes in direction of sound propagation can also be accommodated.

The boom has the advantage of being controlled by an operator who is very well placed to view the action and to avoid as far as possible creating problems for other aspects of the production. The boom can be raised to clear the picture for long-shots and dropped in for close-ups so that the dual requirements of keeping the microphone out of the picture and matching visual and aural perspective are achieved in the same operation.

Disadvantages
Problems arise in the use of booms:

1. When artists work close to the walls of the set.
2. Where the lighting is in line with the camera shots (see page 46).
3. Where there are pillars or arches or other tall objects in the foreground.
4. When the nature of the programme material or the need to achieve separation requires a close microphone technique regardless of the headroom in the picture.
5. When longshots, or shots with excessive headroom alternate rapidly so that there is insufficient time to adjust the boom.

1

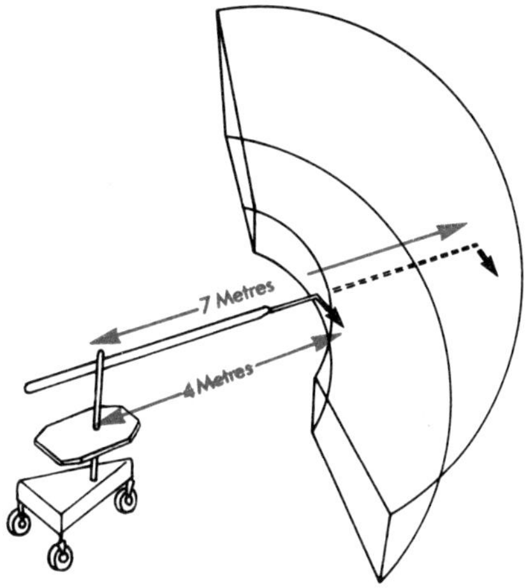

2

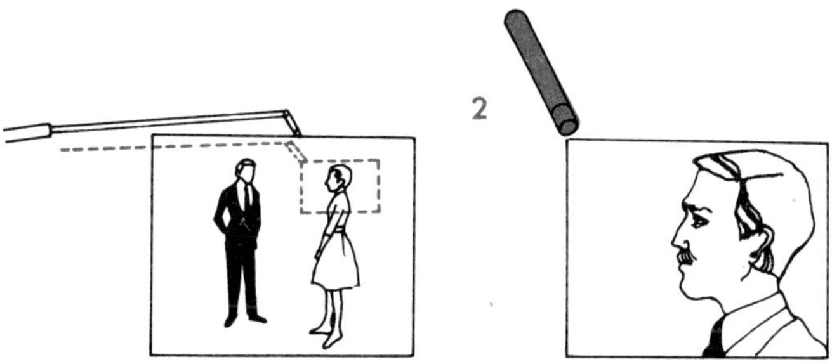

3

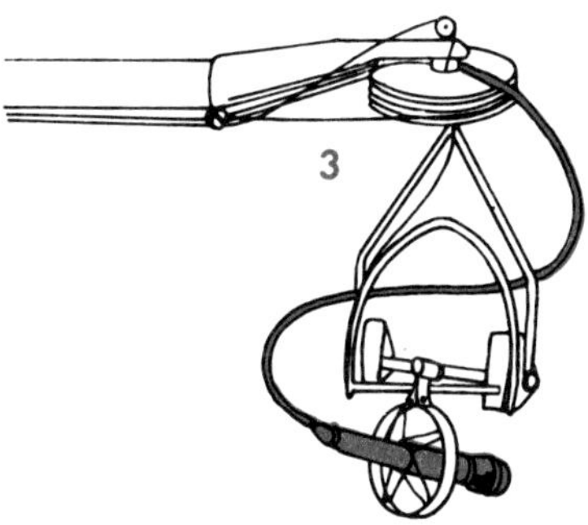

MICROPHONE BOOM

Capabilities and limitations

1, A full sized boom is able to cover sound within an arc of about 180° in the horizontal plane and 30° in the vertical direction with a range of radius between 4 and 7 metres, assuming the artist is about one metre in front of the microphone. 2, Held above the action the boom can be quickly raised or lowered to suit the headroom in the camera shot and to match perspective. 3, Detail of boom microphone suspension.

Boom operating requires a thorough knowledge of microphone technique and characteristics, visual imagination and a good memory.

Boom Operating

The boom operator's task is to maintain his microphone in as nearly as possible the ideal position and pointing in the optimum direction throughout the action, while causing the minimum restriction and inconvenience to the other aspects of the production.

The boom operator is able to receive directions via his headphones from the sound supervisor but he should not have to rely on that. He should be able to assess for himself as accurately as possible the relative levels of the sounds he is picking up and the requirement in terms of aural characteristic e.g. perspective etc. to match with the camera angles and apparent viewing distances from instant to instant.

Operator's responsibilities

The boom operator must have a keen programme sense and the ability to memorise dialogue. Although he can be provided with a script and shot-list which he clips on a special board, he will not be able to read it while he is involved in a sequence because to look away from the boom would probably result in the microphone coming into shot and could even be dangerous, for example if a seated actor suddenly rose to his feet while the microphone was above his head.

The boom operator must assess the position and direction to point his microphone not only for optimum sound pick-up but also from the point of view of providing maximum discrimination against unwanted sources of sound, i.e. orchestral accompaniment (unwanted on the boom) or PA loudspeakers, etc. To this end he may be more concerned with pointing the dead side of the microphone towards the unwanted source than in pointing the live side directly at the artist.

The boom operator is provided with a reverse talk-back microphone with which, by pressing a key, he can communicate with the sound supervisor. In this way he is able to keep the supervisor informed of the situation and problems that arise on the studio floor, especially during rehearsal, which his commanding position makes him particularly well favoured to observe.

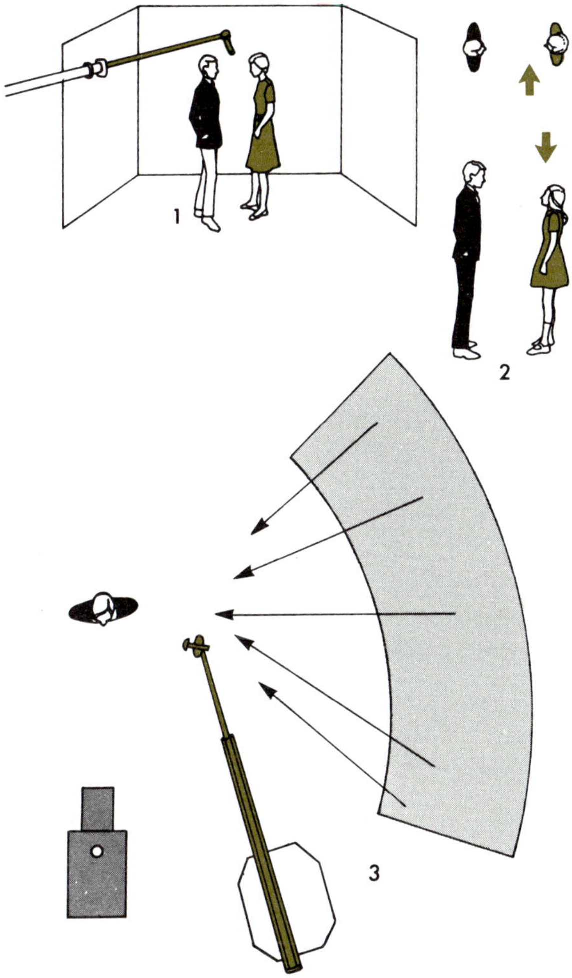

MICROPHONE BOOM

Operation

1,The usual position for the boom in a dialogue situation is above the level of the actors' heads and slightly downstage of the action (i.e. toward the cameras). The actual height and distance depend on the camera shots and perspective requirements or the need to discriminate against unwanted sound. 2, The microphone should be positioned and angled to favour the weakest voice or shortest person. The figure gives an indication of the position in plan and elevation. 3, It may be more important for the boom operator to point the dead side of the boom toward the main source of unwanted sound than directly toward the artist.

Television sound operations and lighting are interrelated and require mutual understanding.

Basic Lighting Requirements

One of the most difficult problems associated with boom operating is to avoid making shadows, either on the artists or on the scenery, which can be seen in the picture. Boom operators should be familiar with the basic principles of production lighting and be able to recognise the various light sources and assess the risks involved.

Lamp functions

Very many lamps are employed in most television productions, mainly because of the need to light for a number of different areas and camera angles simultaneously. If the situation in any one position is analysed it will be found that four, or possibly five, basic arrangements are involved.

1. *Key Light.* A hard (i.e. point) source of light from a lens spotlight intended to provide modelling and show texture and contour by creating shadows.

2. *Filler.* A soft (diffuse) source of fairly large area usually obtained by directing a number of light sources through a diffuser or by reflecting light from point sources, which are not directly visible, via a large reflector. These sources are intended to provide light in the shadow areas left by the key light to show detail and give artistic balance.

3. *Back Light.* A light shone from above and behind the artist to put shine on the hair and rim the outline to promote a three dimensional effect.

4. *Effects Light.* Lighting for the set or backing to the artists. It is intended to bring out the architectural features and promote a sense of realism (e.g. sunlight streaming through a window or lamplight at night). Modern light entertainment colour productions often rely upon the use of colour lighting on a white set to effect changes of scene or mood.

5. *Camera Headlight.* A light sometimes attached to the camera to soften the shadows under the eyebrows and chin and create catchlights in the eyes in close-ups.

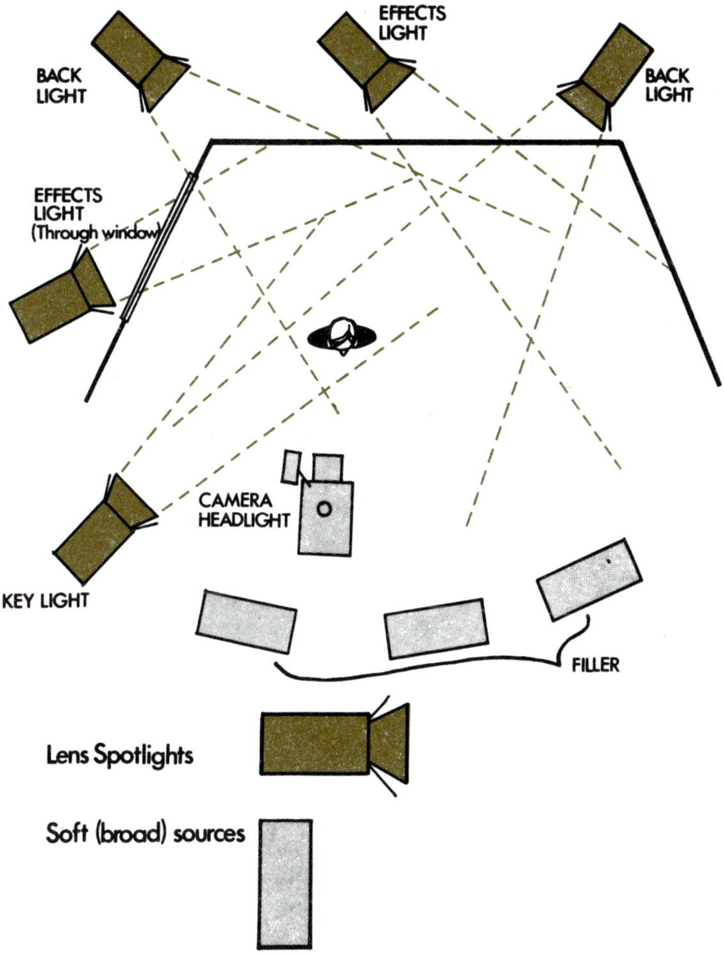

LIGHTING

Basic arrangements
A typical set in plan view showing basic lighting arrangements. Many more lamps may be required if a number of acting areas and camera viewpoints is involved, but the basic lighting pattern in each position will be much the same.

The Boom and Lighting

The preceding section shows that most shadow problems are caused by the key lights. As the key light is a point source, any object such as boom microphone placed in its path will throw a distinct shadow. For good portraiture the key light is required to subtend only a small angle to the direction of the artist's face.

If the boom and key light are in line with each other a shadow appears on the artist. The boom must therefore approach from the opposite side of the artist to the key light. When seen in long-shot the artist must work sufficiently far from the scenery so that the shadow of the boom is lost on the floor or on some part of the set that is not in shot. Luckily most situations involve dialogue between two or more people facing each other across the set. In this case the key lighting can come from the sides or even to the rear of the set so that shadow problems are reduced.

Soloists in light entertainment usually require frontal key lights but tend to work with hand-held or lavalier microphones.

Multi key lights

Continuous-take production can involve the use of a number of acting areas and therefore a number of key lights on each set. Boom operators should study the lighting arrangement before positioning the boom and lighting directors must take account of boom position before designing their lighting plot.

Heat problems

Sound operators should bear in mind that many of the lamps used in television radiate a considerable amount of heat as well as light. They should be careful not to position microphones, especially electrostatic ones, in close proximity to the lights.

Lamp sing

Light sources that are controlled by electronic dimmers (e.g. thyristors) sometimes radiate quite a loud high-pitched sound due to movement of their filaments. This can be accentuated in certain types of housing such as 'scoops' (a bowl-shaped soft source) which tend to have a marked focusing effect.

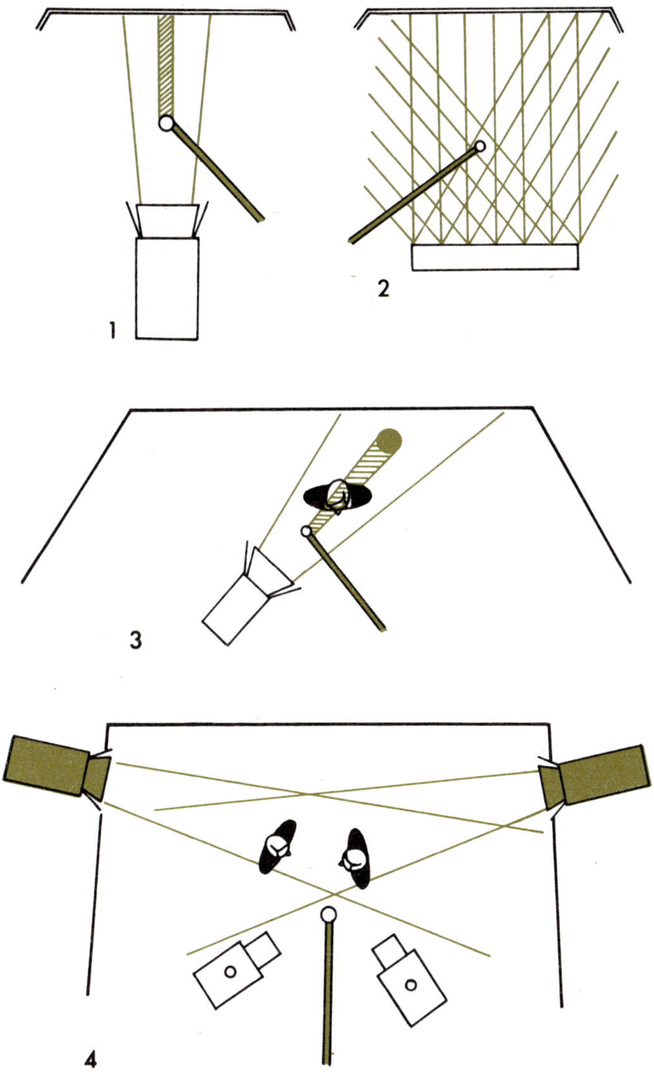

LIGHTING

Avoiding boom shadows

1, A point source of light throws a sharply defined shadow of the boom microphone on to the set. 2, A broad source of light has many overlapping beams in which the individual shadows are cancelled out. 3, When the artist is positioned well forward from the set the shadow of the microphone is thrown on the floor where it cannot be seen. 4, A method of using three-quarter back lights for cross-keying a dialogue sequence. As there is no frontal key light there is no shadow for the boom.

The Boom and Camera Movement

It is only natural for television directors, whose main preoccupation is with directing the cameras to look upon the microphone boom as a necessary nuisance and to be somewhat apprehensive if it is suggested at a planning meeting that several are required even on one set. In fact modern booms are able to track or crab and are almost as manoeuvreable as cameras. They also have a considerable reach and there is seldom much danger of their restricting camera movement.

Cable arrangements

The essence of the operation is proper planning in relation to the cameras and a well organised system of cabling. The best way to provide flexibility of movement for the boom and at the same time prevent the boom cables from becoming entangled with the camera cables is usually to sling the boom cables from above, preferably by means of special suspension points which can be lowered from the roof grid of the studio. These can consist of simple slung cables and connection position or sophisticated devices consisting of motor-operated telescopic tubes with connection points mounted at their base.

The connections available should include the microphone lead (possibly with 'phantom' power connection for powered microphones), talkback and reverse talkback lines, foldback circuits (for use when loudspeakers are to be mounted on the boom) and a video cable to enable the boom operator to be provided with a monitor. This can be of considerable assistance where wide variations of picture angle are involved or the boom operator is not able to see the cameras clearly because of some obstruction.

In a large studio four slinging positions would be required, arranged approximately as suggested in the diagram. At the boom end the cable should pass through a 'fly away' device normally consisting of a metal spiral into which it can be threaded without disconnection unless the suspension system is fitted with a cable counterweight. A few metres of the cable is normally coiled up on the platform of the boom to allow for travel, the slack being taken up and re-coiled as the boom approaches its suspension point.

Boom positioning

The normal position for a boom to work on a set is between, but slightly behind the cameras. If the cameras are mounted on pedestals it is possible for the boom to work over the top, but if any sort of crane is involved, and in particular a swinging crane, the boom has to work to the side (possibly even to the side of the picture frame).

Some of the largest cranes are provided with spigots to accept microphone boom arms so that the operator can ride with the camera.

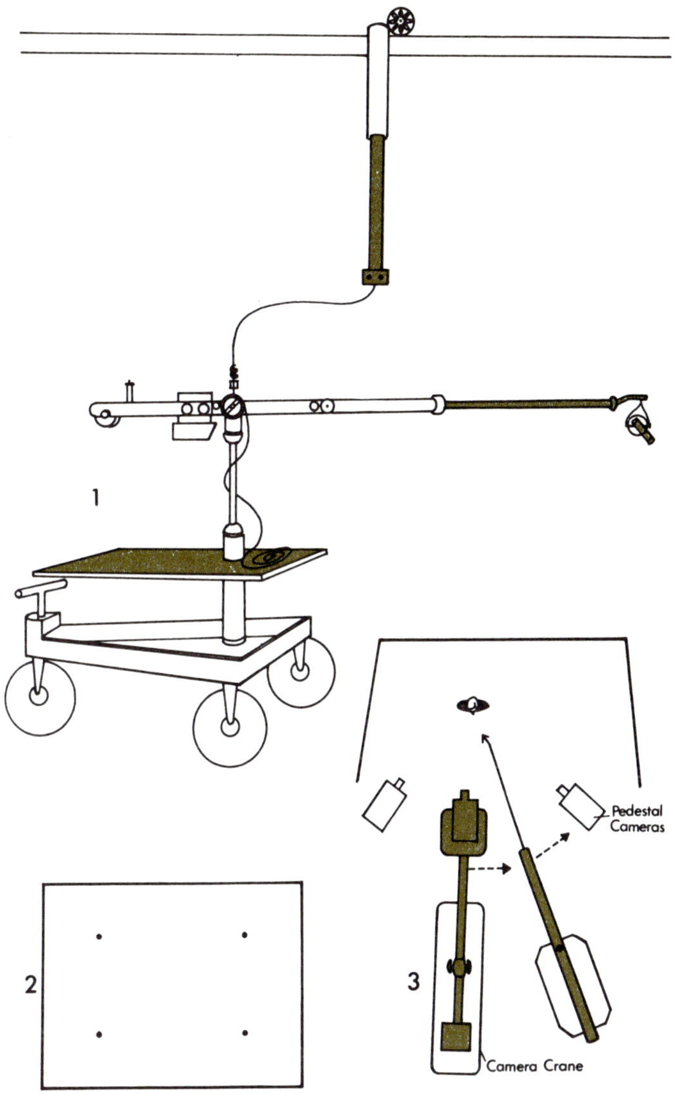

BOOM AND CAMERA

Allowing for camera movement

1, Electrically operated telescopic hoist for suspending microphone cable termination from lighting grid. 2, Plan view of large studio showing the general disposition of four boom termination hoists. 3, Boom working alongside cameras, one of which is a crane with a jib that can swing from side to side and up and down. If the crane has to swing toward the boom while elevated, the boom will have to swing also. This is a point that should be taken into consideration at the planning stage as it might require the boom to be repositioned or to have another one on the other side of the crane to take over. If the boom has to execute a wide swing, the 'pram' must be angled so that the operator does not 'run out of platform'.

Planning for the Boom

Planning for the boom centres around two types of scale drawing. The studio plan and the elevations. Sometimes models are also used.

The studio plan

The studio plan consists of a 1:50 scale plan of the studio, usually marked out in approximately 6 mm squares, each to represent 30 cms. The outline of scenery is marked on the plan to scale and the starting positions for the cameras and booms are marked by means of a special stencil. The positioning of the cameras and booms is decided at a planning meeting prior to the production day attended by the director, sound supervisor, lighting director, set designer and any other specialists concerned at that stage in the proceedings. It is explained that the marked position can give only an approximate indication of the starting position because both cameras and booms are mobile. No matter how careful the planning, shots develop and positions become modified as a result of rehearsal experience. Nevertheless it is necessary to have a pretty firm basis from which to work.

During the course of a production the cameras and booms will probably be required to work in a variety of different areas and these are identified on the plan by the camera numbers or boom letters and position numbers (i.e. cam. 1 pos. A; boom A pos. 2, etc.).

Part of the art of camera planning concerns arranging the cabling so that the cables do not get tied up in knots as the cameras move about and this goes for the booms too if the cables are not slung.

The elevations

The elevations are detail drawings of the vertical appearance of the sets. This can be very useful at the planning stage in judging heights and particularly if obstacles are to be encountered such as roofs, pillars or arches etc.

When very complicated structures are involved it is usual to make models of the sets from balsa wood or cardboard to the same scale as the plan and these can be invaluable in helping to assess the optimum position for microphone coverage and the hazards of boom operations.

The actual detail of boom planning is bound up with the optimum sound requirements. This depends to a large extent on the nature of the production and is dealt with in detail in later sections.

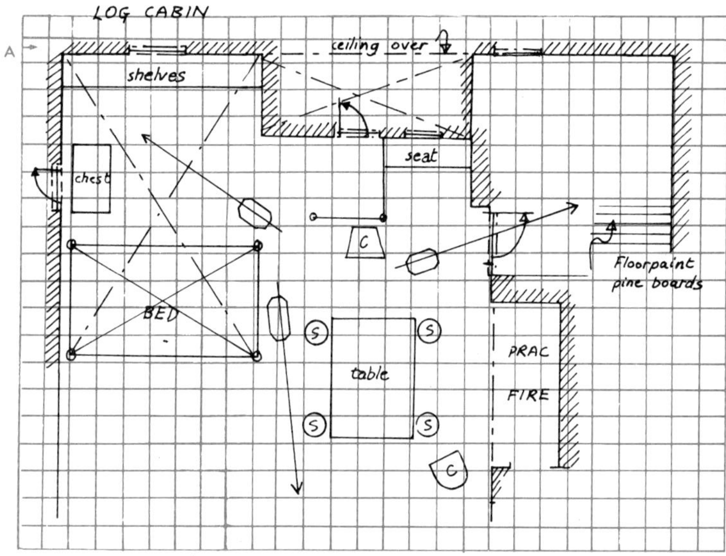

1

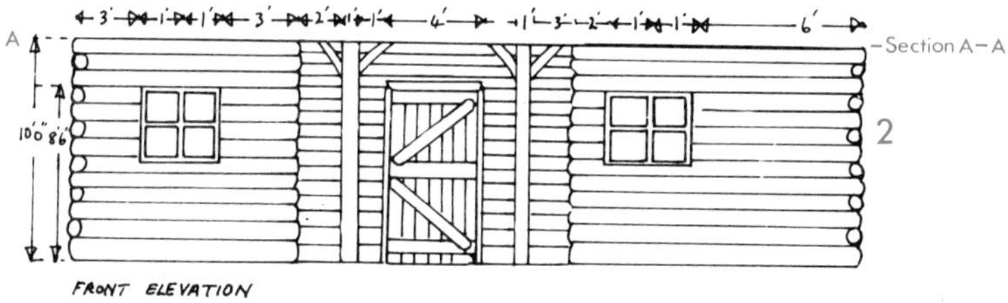

2

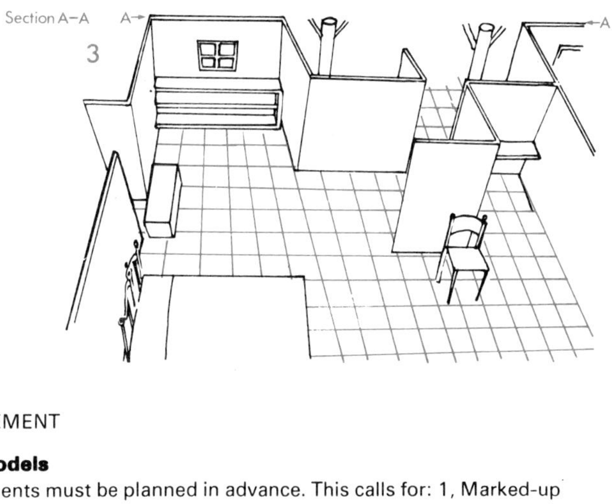

3

BOOM MOVEMENT

Plans and models

Boom movements must be planned in advance. This calls for: 1, Marked-up studio floor plan, showing disposition of booms. 2, Elevation, indicating the vertical appearance of the sets. 3, Model of complex item of scenery.

Production Talkback Arrangements

Television production is, above all, a team operation which relies heavily on good communications. Continuous-take or live production would not be possible unless all the operators concerned were able to hear the director simultaneously. For this reason, all camera and sound staff and any other operators working on the studio floor who are required to take cues, wear headphones (often called 'cans' because of the can-like appearance of the metal ones used for many years in radio and television).

Director's talkback

The main talkback source is the director's (or producer's) microphone. This is normally live throughout the performance. It is fed to all the operators working on the studio floor as mentioned above and also to any other areas concerned with the production. These would include the lighting and sound control rooms, (if separate from the production control room) any telecine and videotape channels associated with the production, any remote sources that may be involved and even to loudspeakers in the artists' dressing rooms.

The producer's talkback must be clear and of reasonably consistent volume, so directional microphones are usually employed (to discriminate against the programme loudspeakers and other ambient sounds) and a volume limiter is incorporated. The director is also provided with loudspeaker talkback into the studio, controlled by a key, but seldom uses it because it tends to cause confusion.

Floor manager's talkback

The director's directions are relayed to the artists by the floor manager who is therefore required to be very mobile. He normally hears the director by means of a low power (about 500mW) transmitter and a small receiver which he carries in his pocket or a sling harness. A reverse talkback transmitter can be provided for the floor manager to speak back to the producer but this facility is not often required as there are usually programme microphones, e.g. a boom available in the studio.

Camera talkback

The director's talkback is fed to all the cameramen's headphones via the camera cable. The cameramen have the facility to listen also to programme sound. This can be mixed with the director's talkback to the required proportion to augment that inevitably picked up from the loudspeakers in the production control room. In some systems the headphones are split to provide talkback in one ear and programme in the other.

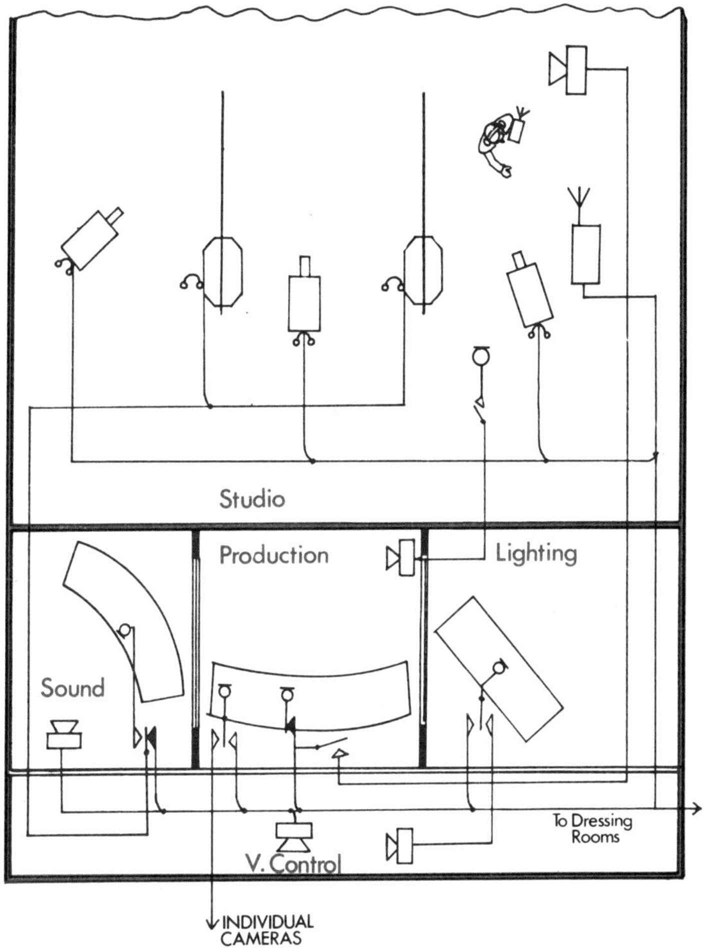

PRODUCTION TALKBACK

Typical studio talkback arrangement
The producer's talkback is the only circuit that is normally connected to everybody continuously. That of the sound, lighting, technical manager and vision control is directed to the respective personnel via switches.

The sound supervisor can override the producer's talkback to give directions to the sound operators.

Sound and Lighting Talkback

Just as the director uses his talkback to direct the movement of cameras about the studio floor so the sound supervisor and lighting director direct the sound floor staff and electricians over separate talkback systems.

Sound talkback

It is not possible for the operators at the microphone to judge the effect of their work. They need guidance during rehearsal, although good boom operators cultivate such a good memory that they should manage with little direction during transmission. Nevertheless they need to be told when changeovers between the boom have been completed and they are released to go to their next position. As they seldom have time to read a script it is a good idea to warn them of important cues and moves.

The sound operators on the studio floor hear the same talkback as the cameramen until the sound supervisor wishes to speak to them. He then operates a key, or foot switch, which exchanges his microphone output for the director's. This also gives the sound operators a short burst of tone as an indication that the sound supervisor is intercepting the circuit.

Reverse talkback

The boom operator is provided with a reverse talkback system consisting of a key-operated microphone mounted on the boom. He uses this to warn the sound supervisor of any problems that arise in the studio that he may not be able to see and to cue him as to the best moment to change over from the output of one boom to another in a take-over situation.

Lighting talkback

A similar arrangement also exists for the vision control staff and the lighting electricians. They normally listen to the director's talkback until it is intercepted by the lighting director. Lighting directors normally have available small two-way radios working between lighting control room and studio floor to facilitate lamp adjustment during setting periods and rehearsal.

Wall points

All three talkback circuits (producer, sound and vision) are available at all wall points and suspended boom connections.

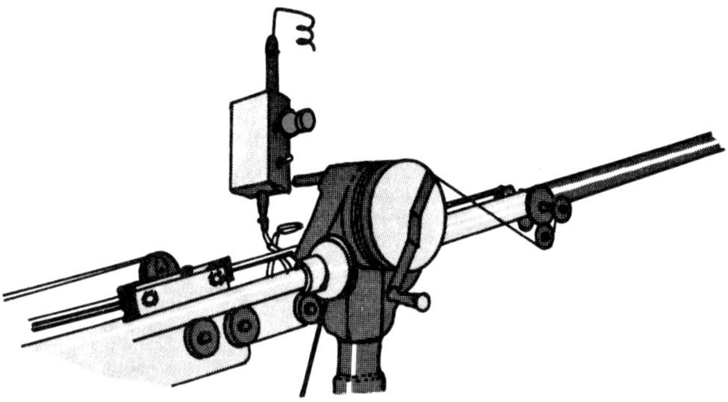

SOUND AND LIGHTING TALKBACK

Boom mic and walkie talkie
The boom operator can be provided with a microphone and switch box mounted on the boom *(top).* If he wishes to talk to the sound supervisor during the action he has to approach very close to the microphone and use a very quiet voice. Alternatively he can be provided with headphones which incorporate a small close talking microphone. A typical walkie-talkie communication set in use for setting lamps *(bottom).*

Sound Control

Most reproducing equipment has a volume control, so neither the artist nor the sound operator can have any say over the actual volume at which their material is finally reproduced. They can merely affect the *relative* loudness of the various parts of the programme and its individual elements.

Basic requirements

The reasons for volume control were outlined briefly on page 14. At this stage it is worth considering the matter in more detail. There are three basic requirements:

1. To keep within the dynamic range of the system. In most cases, such as recording and live transmission, the upper limit is set by the onset of overload distortion and the lower by the threshold of intrusive noise.
2. To adjust the scale of volume range to suit the listening conditions and environment. Most types of programme material, from music to drama involve a range of volume that would be quite unacceptable in average domestic circumstances.
3. To ensure that items produced in different areas and at different times match each other and in broadcasting maintain a smooth balance between successive programmes.

Thus, there are two aspects of volume control — the technical and the artistic. As we are not able to measure absolute sound levels with our ears and have a poor memory even for relative levels we need a meter (as described on page 62), and a set of rules regarding the proper level for each type of programme material if we are to keep within the technical limits and provide a balance between successive items or programmes. But the meter does not help us to make artistic judgments or even assess loudness, which is largely subjective.

Use of loudspeakers

The only way to effective programme control is by means of a good quality loudspeaker, making use of a meter to 'calibrate' the ear in the way that a car driver judges his speed without watching the speedometer all the time.

The importance of listening levels

Practically all operators professionally engaged in sound control listen at very high volume. This is necessary to achieve the right degree of concentration and because otherwise there is a tendency to 'wind it up' out of sheer enthusiasm. Unfortunately this can give a very unreal picture of what the listener will hear and can lead to a misleading balance, particularly as between vocal and accompaniment or speech and sound effects. It is therefore advisable to 'dim' the loudspeaker from time to time and get used to making allowances for the difference.

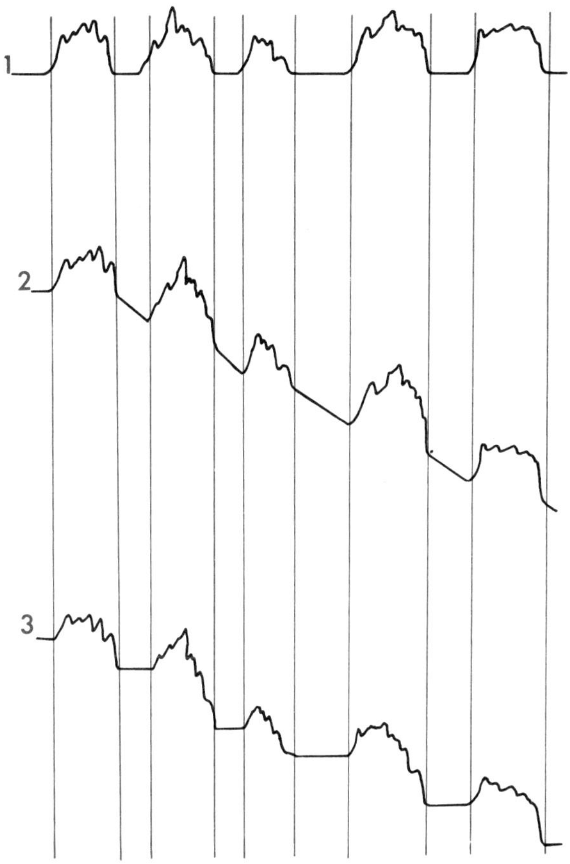

SOUND CONTROL

Speech modulation
1, Most sources of sound are continually varying and interrupted in character and a fade which continues through the gaps in the material would result in a succession of steps in the output.

Unobtrusive control
2, If the intention is for the operation to be as inconspicuous as possible, which is the case for most types of material and particularly music, the action should occur in the gaps in the material or during rapid fluctuation of level.

Control for dramatic effect
3, If it is intended to make an obvious smooth fade for dramatic effect (eg. a slow fade on dialogue is the conventional method of suggesting a change of scene in radio drama) movement of the control must occur only during the material, preferably during sustained phrases.

Programme control should be exercised only at source and should affect only those items that require it.

The Points of Control

The previous topic discussed the need for control. Now we consider the points at which programme control is exercised.

Professional sound control desks come in all sizes and degrees of complexity to handle any number of channels from less than six to more than forty. On the larger control desks, programme control can usually be exercised at various alternative positions in the chain.

Pre-set (or 'balance') attenuator

For television sound the range of levels applied to the channels can be enormous, varying perhaps from whispered dialogue picked up by a microphone at an excessive distance (in order to be out of shot) to a shout or a musical instrument at very close range. This can represent a variation in microphone level possibly of up to 60 dB (i.e. 1,000,000:1). In order to enable the apparatus to cope with such a range, television desks are usually equipped with pre-set controls in addition to the channel controls. These are usually small variable attenuators by which the gain of each channel can be adjusted.

Channel fader

The whole essence of effective programme control is that it should affect only those elements of the signal that render it necessary (e.g. imagine the sound of a whole orchestra being diminished to accommodate a beat of a drum). Thus the most important controls are usually the individual channel faders.

Group fader

Although ideally control should be exercised on the individual channels there are circumstances when the operator has no time, or insufficient fingers, to adjust all the required number of controls simultaneously. It is usually possible to select a number of sources that can be conveniently grouped together. These might include, for example, several microphones in use for a large choir, or a section of an orchestra, or even the whole orchestra. The object being to group together sources that might have to be balanced as a section against other sources.

Main fader

Last in the sound desk comes the main fader (or main faders, if more than one output is provided). The main function of this control in modern practice is to enable a balanced mix of all required sources to be faded in and out simultaneously and to set a convenient overall level. For this reason it is seldom altered throughout the performance and can be considered almost a pre-set device.

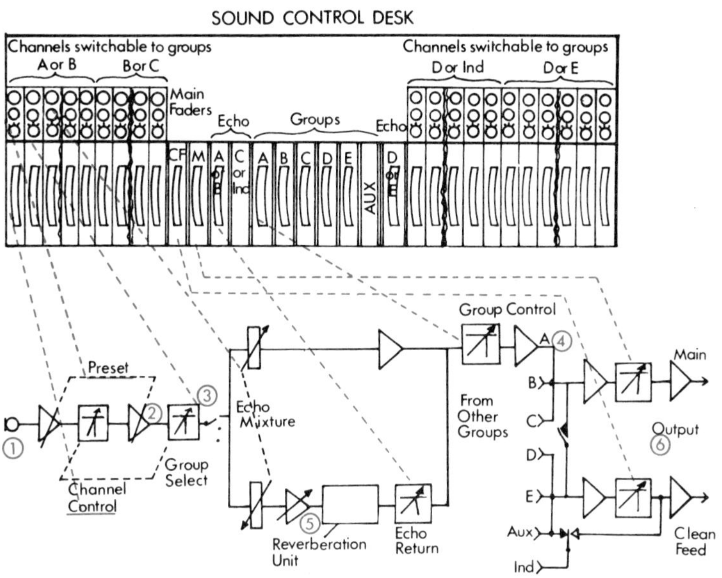

POINTS OF CONTROL

Control point levels
Typical programme chain of modern sound mixing desk. Average levels at various control points are: 1, Input — 70 dB. Pre-set amplifier controls (set to 10 dB below full gain) — 18 dB. 3, Channel fader (set to 10 dB attenuation) — 18 dB. 4, Group control (set to 10dB attenuation) — 18 dB. 5, Input to reverberation unit (at 50/50 mixture) — 18 dB. 6, Output (with main control set to 4 dB attenuation) zero level. The maximum gain of the desk with all controls flat out is approx 105 dB. Gain with controls in the above positions for optimum signal-to-noise ratio and minimum risk of overload is 70 dB.

The Sound Control Desk

The sound control desk is a device for collecting the output from a number of sound sources, controlling them and processing and combining them into a single output or multiple outputs for recording, reproduction, or transmission.

Programme volume control

The points at which sound volume can be controlled in television sound control desks were discussed in the previous topic. The number of channels required for a sound desk depends on the maximum number of sources and the flexibility of the grouping arrangements. The ancillary facilities for a control desk to be used for complex productions would include the following:

1 *Pre-hear.* The ability to listen to each channel before fading it up. A useful feature if connections have been changed or microphones repositioned during transmission.

2 *Prompt cut.* A relay that interrupts the dialogue microphone output, but not the gram/tape output so that by pressing a button in the studio prompts can be given to the artists in live drama without them being broadcast.

3 *Foldback.* A method of playing back music, effects, or other sources such as remotes or pre-recorded material into the studio over portable loudspeakers for the benefit of the artists (see pages 148 and 154).

4 *P.A.* The ability to feed the output of some microphones and other sources to a loudspeaker system for the benefit of an audience. There are various points in the chain where these feeds can be obtained (see page 84).

5 *Echo.* Complex productions may require up to three different characters of artificial reverberation simultaneously. Separate echo circuits are therefore provided for each group (this is necessary anyway to avoid bypassing the group control). They can be selected to echo room or reverberation plates as required. Each channel has a control to set the level of input to the echo device (and hence the individual direct/echo ratio) and the outputs are returned through 'echo' faders by which the amount of echo used can be controlled.

Frequency response control

Most modern sound desks are provided with frequency response selection controls or provision for their insertion in channels or groups (see page 66).

Multiway clean feed

For providing a number of outside sources contributing to a hook-up with every output except their own (see page 92).

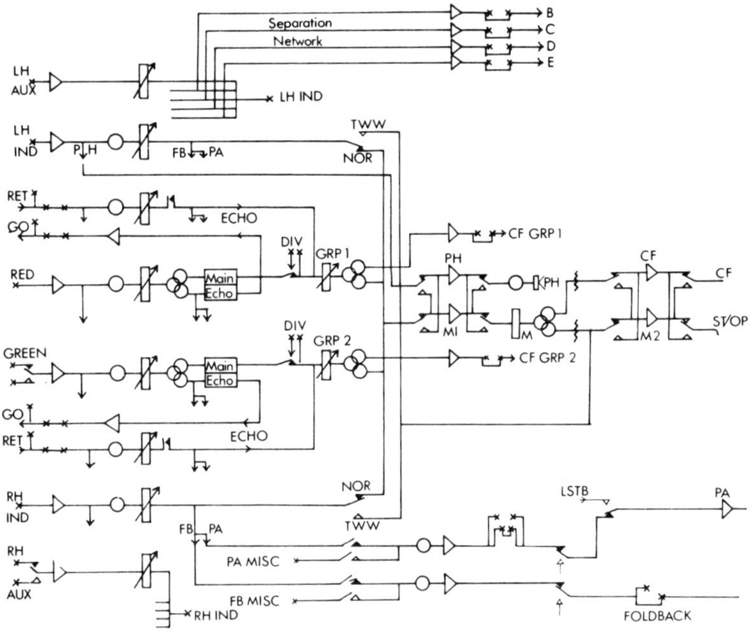

SOUND OPERATIONS

Sound control desk

Basic schematic of a simple sound control desk showing alternative facilities to which sources can be selected. The left and right hand auxiliary connections provide the facility of 'multi-way working'' ie reverse feeds to a number of contributors. The left and right hand independent channels offer the alternative of 'normal' facilities (except echo) or 'two-way-working' in which the output bypasses the main fader so that remote sources, connected to independent channels, can be provided (by means of the CF output) with every contributor's output except their own.

The main groups (here marked red and green) each have a separate echo system.

All channels are provided with pre-hear, foldback and PA facilities. The pre-hear and clean feed amplifiers also provide an emergency change over facility for the main output amplifiers.

Programme Meters

In a previous section we discussed the need for volume regulation and the requirement to keep the volume within certain prescribed limits, both for practical and for artistic purposes. Some form of meter must be provided in order to set volume levels and assist in the assessment of volume range. There are two types of programme meter in professional use, the volume unit meter and the peak programme meter.

The volume unit meter is the simpler of the two. It consists of a rectifier and moving coil meter which has two scales, one above the other. One scale is calibrated in decibels, indicates volume level and is intended for use only with steady tone. The other is calibrated in volume units numbered in black from 0 to 100 over about $\frac{2}{3}$ of its range. The last part of this scale is uncalibrated, and marked in red. This scale represents percentage use of the channel and is intended for use on programme. The needle has specially designed ballistics intended to integrate rapid variation but even so it can be difficult to read on rapidly fluctuating material. Impulsive sounds can cause the needle to overshoot and give a higher reading than sustained sounds of a similar volume.

The peak programme meter has a needle which is actuated by circuitry that allows it to rise very rapidly and fall comparitively slowly so that rapid fluctuations are smoothed out and the meter is always easy to read. The meter can rise to full scale deflection in approximately four milli-seconds and recover to 'one' in approximately three seconds.

Two types of peak programme meters are in use in television. The standard meter has white figures painted on a black background. The other is an optical meter in which the indication is given by a beam of light (incorporating a hair line cursor) projected by a movable mirror on to a ground-glass screen. Both types of meter have scales calibrated in decibels with the figures from 1 to 7. There is also a zero mark and a full scale deflection mark. The spaces in the working part of the scale between 2 and 7 represent 4dB each. Normal practice is to line up equipment with tone at some middle frequency such as 1 kHz to a level of 8dB below the maximum — represented by the figure 4 on the peak programme meter. Full modulation is represented by 6.

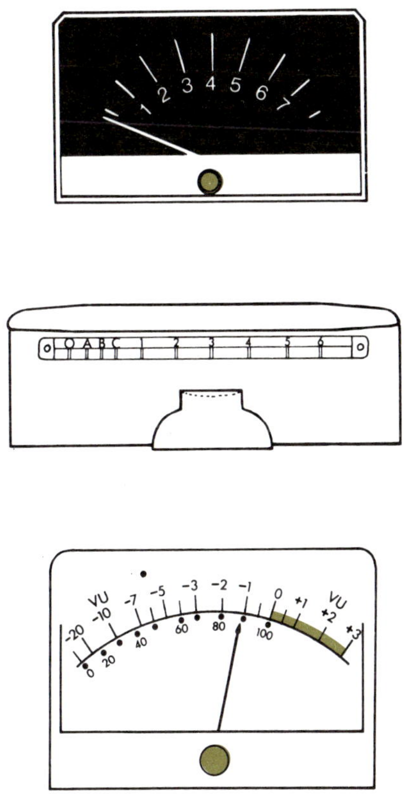

Types of meter
Top: Peak programme meter (the figures are in white on a black background to reduce eye strain).
Middle: Optical programme meter. These meters are intended to be placed just beside television viewing monitors so that operators who have to study the screen and programme meter fairly consistently do not have to keep adjusting the focus of their eyes.
Bottom: Volume unit meter.

Automatic Control

No automatic device can provide the necessary aesthetic judgment and finesse for effective programme sound control. Nevertheless automatic control systems (known as compressors and limiters) have considerable application in sound operational practice to assist in the 'taming' of wildly varying and unpredictable sources and as a long stop to protect equipment from possible overmodulation.

The limiter

The limiter is a device through which programmes can be passed without either gain or alteration of the signal until a critical value is reached. If the input signal rises above this value (usually called the onset point) the gain of the system is automatically reduced (below unity) to such an extent that the output is not allowed to rise significantly above the limiting value. This limiting action is caused by reduction of the amplifier gain, not merely by cutting off the peaks of the waveform which would be 'peak chopping' and would result in very severe distortion. Limiters are commonly employed at transmitters to protect the circuits from overload and, in the case of frequency-modulated transmitters, to prevent over-deviation into adjacent channels.

The compressor

The compressor is similar to the limiter, in that above the onset point the gain of the system is reduced, but less dramatically, so that an increase in input level above this value produces an increase in the output but to a lesser degree. The extent of the gain reduction is usually adjustable and is called the *compression ratio*. Other controls usually available on compressors are:

1. The *onset* or *threshold point* (the value at which the action commences).
2. The *attack time*, being the time required for the action to take effect following a sudden increase in the level.
3. The *recovery time*, which is approximately the time required for the gain to be restored to unity after the high-level signal has been removed.

The attack time is set as a compromise between too fast, which can alter the shape of the waveform and thereby cause distortion, and too slow with consequent overshoot.

Recovery time is usually set to make the variations of gain as inconspicuous as possible without allowing isolated peaks to depress the level for too long.

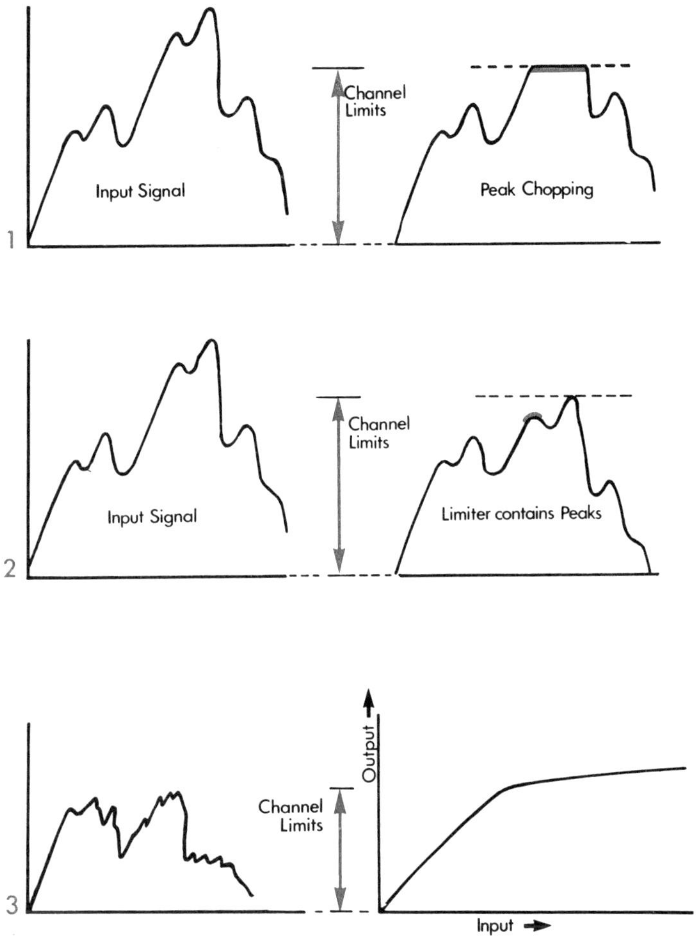

AUTOMATIC CONTROL

Compressors and limiters
1, Effect of peak chopping with severe distortion. 2, A limiter should cause gain reduction to contain the peaks of the waveform. 3, Input/output curve for a limiter showing the 'knee' or onset point of compression.

Frequency Response Control

While it is true that much of the effort of sound engineers and operators goes into promoting high quality reproduction there are instances when to achieve the desired effect requires the controlled application of various forms of distortion.

Filters

Various types of frequency/volume shaping filter are in regular use in television sound. The simplest and probably the most severe is the programme effects unit used for introducing deliberate and obvious distortions as for telephone effects (see page 116).

Response selection amplifier (Equalizer)

Probably the most useful type of filter is the response selection amplifier incorporated in all modern sound control desks — in some cases one for every channel. RSA's provide the ability to accentuate or diminish the treble and bass response usually by varying the slope of the curve about a middle frequency — about 1 kHz. These filters are intended to correct for non-linearity in the source (possibly due to difficult circumstances as when a microphone has to be used underneath thick clothing) or to provide special emphasis for dramatic effect. Also included in the RSA are a series of 'presence filters'. These provide a range of quite sharp peaks, usually of the order of 3 or 6 dB, at a choice of several pre-set frequencies in the region 2 to 6 kHz. They are called 'presence' filters because the peaks are usually centred on the frequencies that the ear associates with closeness, particularly the sibilants in speech, which are soon lost with distance. Presence filters can thus help to compensate for a too-distant microphone technique or help a vocal to 'punch through' a loud accompaniment.

The octave filter (Graphic equalizer)

A more subtle form of frequency response control is the octave filter (sometimes called a shaping filter). The octave filter provides control over almost the whole sound spectrum by dividing it into nine octave bands centred approximately on 40 Hz, 80Hz, 160Hz, 320Hz, 640Hz, 1·28 kHz, 2·5 kHz, 5·12 kHz and 10·24 kHz. Each can be individually adjusted to the extent of $\pm$ 14 dB.

This can be a most useful device for shaping the frequency response of different microphones or other sources to make them match each other.

Notch filters

A type of filter that has particular application to outside broadcast work is the notch filter. With this device it is possible to put in very steep notches in the frequency response at a choice of frequencies where interference problems are often encountered.

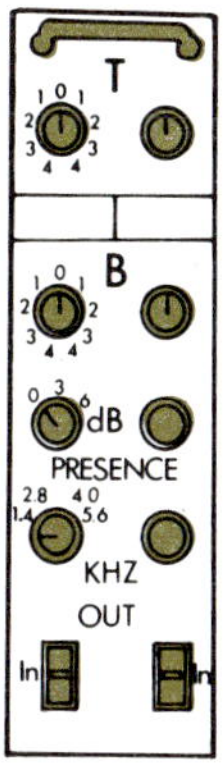

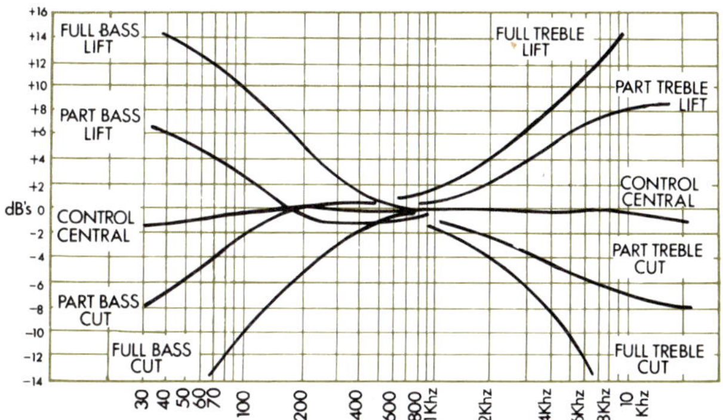

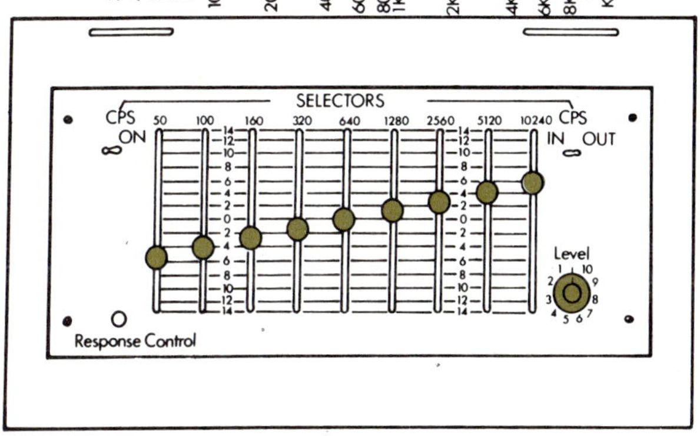

FREQUENCY RESPONSE CONTROL

RSA and octave filter

Top: Appearance and typical controls of response selector amplifier. *Middle:* Graph showing range of control of a typical RSA. *Bottom:* General appearance of octave filter. The sliders each control the response of a separate octave. This arrangement gives an indication of the resulting response curve.

Studio acoustics affect the performance of the artists and the quality of the transmitted sound.

Studio Acoustics

The most important features of the acoustics of a room or studio that affect the sound, and possibly the performance of the artists, are the reverberation time, the shape of the decay curve and the diffusion.

Reverberation time

The reverberation time is a measure of the time a sound takes to die away after the source providing it has stopped. Technically it is the time, in seconds, the sound takes to drop in volume through 60dB (i.e. to one millionth of the original).

The reverberation time is related to the volume of the enclosure and the amount of absorbent material it contains.

Decay curve

Almost as important as the reverberation is the shape of the decay curve i.e. what happens to the sound immediately after the source is cut. A smooth decay curve suggests good diffusion.

Diffusion

Reverberation is caused by sound bouncing around inside a room until its energy has been converted into heat by friction on absorbent surfaces. For good acoustic conditions the reflective and absorbent surfaces should be well distributed so that focusing effects and standing waves are minimised.

Television studio acoustics

The problem about designing the acoustics of a television studio is that, unlike their counterpart in radio or recording; they cannot usually be allocated for one particular purpose. A large television studio may have to accommodate anything from a symphony concert (which sounds best with a reverberation time in excess of two seconds) to an open air sequence in drama which should have no reverberation at all. As it is possible to add artificial reverberation (see page 72) but not subtract it, television studios are made as dead as possible, within the limitations that they should not be too oppressive to work in and the sheer difficulty of finding sufficient unencumbered wall space to treat with absorbent material.

A dead acoustic also helps to damp down noise from all the movement of people and machinery inevitable in the television operation.

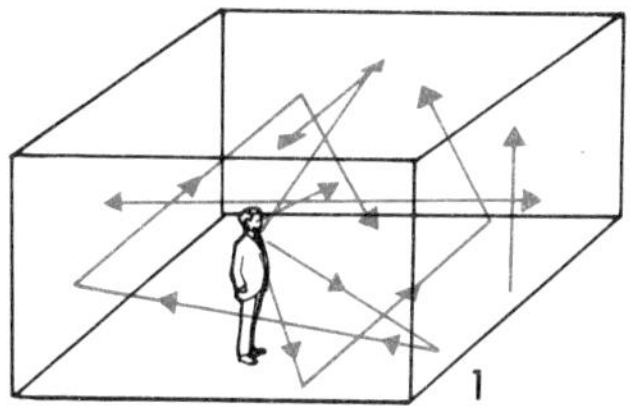

1

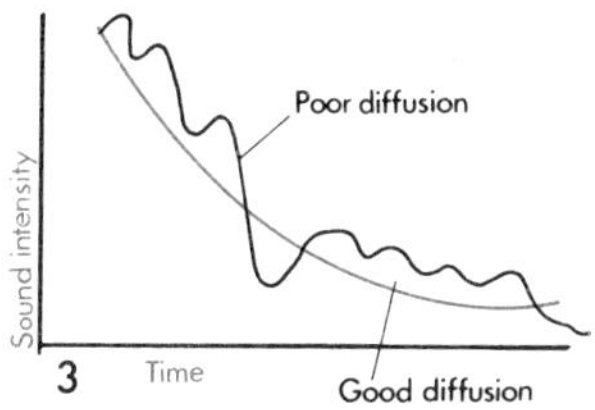

3

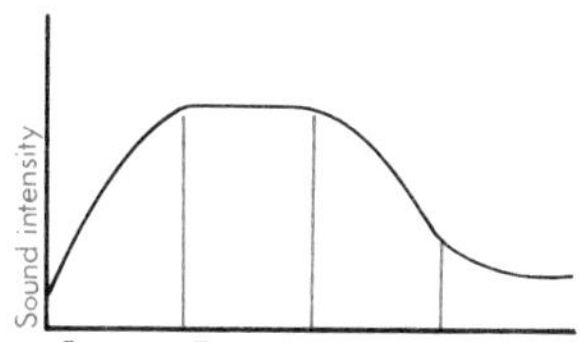

2

4

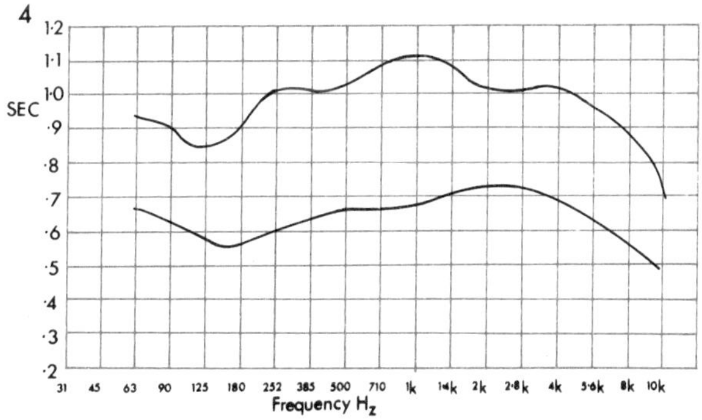

STUDIO ACOUSTICS

Effect of reverberation

1, In an enclosure with reflective walls, sound bounces around in all directions between all the surfaces until it uses up all its energy. 2, When a continuous sound is produced in a reverberant room its intensity builds up as the reflections reinforce each other until a state of equilibrium is reached. When the source of sound is cut, the sound dies away slowly, the rate of decay depending on the reverberation time. 3, The effect that room acoustics have on a sound depends not only on the time the sound takes to die away but also on the smoothness of the decay. A smooth decay suggests good diffusion (well broken up surfaces). 4, Typical reverberation time/frequency curves for a large and a small television studio. They have about half the reverberation time of a typical radio studio of similar size.

The Acoustic Effect of Scenery

No matter how much care has gone into the design of a television studio, its acoustics are likely to be considerably modified when the scenery is brought into it. In fact in the quite usual case where the action takes place in a small parallel-sided set within a very large studio the acoustics of the set have more effect on the sound than those of the studio. Any reflections from the studio that reach the microphone are likely to be too delayed to be of positive use and should be eliminated as much as possible.

Acoustic planning
When planning a production, the sound supervisor should have a clear idea of the effect that the settings are likely to produce, so as to be able to offer advice to the scene designer.

Design consideration
The effect that the settings will have on the sound depends on:
1. The extent to which the material used reflects or absorbs sound.
2. The shape and size of each piece and the angle it makes to the sound source and to the microphone.
3. The possibility of resonance of the material or within any cavities in the framework e.g. rostra.

This is particularly important in the case of music, where sympathetic resonances in the surrounding scenery or rostra can cause unpleasant colouration and spoil the tone of the instruments. Particular care must be taken in the design of rostra for celli, basses and percussion that come into direct contact with the instruments.

Acoustic effect of materials
In general, materials can be divided into the hard and the soft variety. Hard surfaces such as plywood, glass fibre, reinforced resin and aluminium tend to reflect the higher frequencies strongly and, if in fairly large panels, will absorb some bass by acting as a dampened membrane. Curtains, drapes, carpet and soft furnishings are very absorbent at high frequencies and can absorb the bass as well if spaced off from a hard surface to form a cavity.

The importance of shape
The shape of the settings and objects is very important. Look out for focusing effects. It is no good using the directional properties of a microphone to discriminate against noise in front of a reflecting surface that might actually focus it back on to the live face.

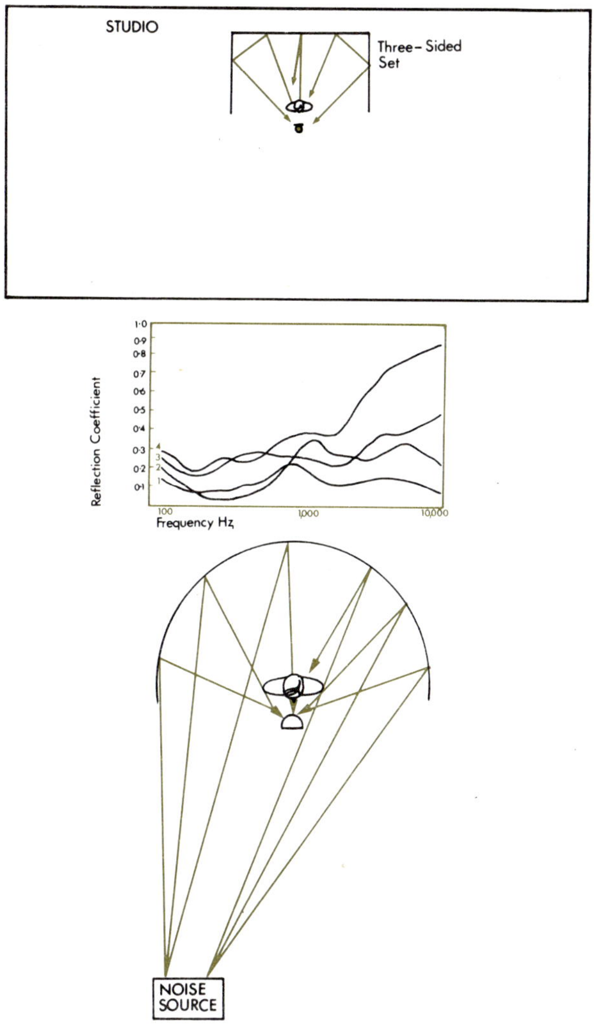

STUDIO ACOUSTICS

Effect of scenery

Top: An indication of the scale of a drama set in a small television studio. The acoustic effect within the set is likely to depend more on the scenery than the acoustics of the studio. Reflections from the studio walls tend to arrive too late to be of practical value. *Middle:* Curves showing the reflective properties of various materials frequently used in the construction of scenery. Note that the reflectivity can vary considerably with frequency. This can cause colouration of the sound. Curve 1, Thick fabric; 2, Painted canvas flat; 3, Wallpaper on plywood; 4, Fibreglass on timber frame (shaped to look like brickwork). *Bottom:* Curvature of set may focus unwanted source of noise on to the live face of the microphone.

Television studios have dead acoustics and artificial reverberation is added when required.

Artificial Reverberation

Although television studios generally need dead acoustics, the sound output should rarely have a dead acoustic. Reverberation is essential for all types of music. It gives warmth and sonority and a sense of power. For drama it can give the effect of distance and solidity of environment. So, where it does not exist, means must be made available to add reverberation artificially.

The reverberation plate

A method of producing artificial reverberation that is particularly suited to television because it offers variable reverberation characteristics to match the varied circumstances, is the reverberation plate. This consists of a steel sheet of about 2 metres by 1 metre suspended vertically under tension within a steel framework by springs attached to all four corners. A moving coil transducer is fixed at a critical point near the centre and a piezo electric contact microphone is mounted near to one edge. When a signal is applied to the transducer a series of very complex vibrations radiate out to the sides and are reflected to and fro between the edges until they die out in the manner of reverberation.

The reverberation time can be adjusted by altering the spacing between the plate and another plate of similar size made of porous material held parallel to it. The closer the plates the more the sound energy is absorbed due to friction of the air particles in the porous material.

Reverberation springs

A reverberation device that is more suitable for mobile use is the spring reverberator. This consists of two sets of two springs of different lengths joined together at the ends. The sound vibrations are induced into one end of the pair of springs in a torsional manner and received at the other end. The energy reflects back and forward between the two ends of the springs until it dies away giving a reverberant affect with a decay time fixed at two seconds.

All artificial reverberation devices sound better if their sound input is delayed by about 100 msec.* This corresponds to the time the natural source would take before bouncing back from the first reflective surface.

* This is usually achieved by recording/reproducing process, the tape (or magnetic disc) speed and head spacing being arranged to give the required delay. In some systems digital memory techniques are used.

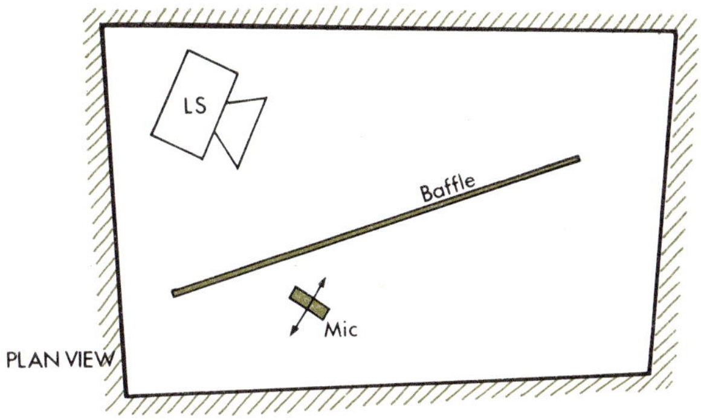

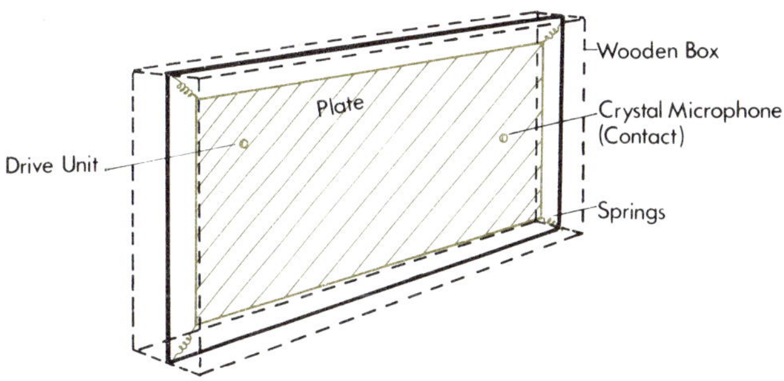

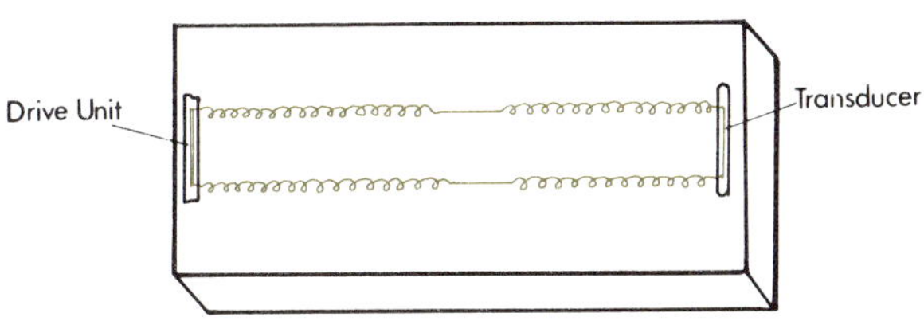

ARTIFICIAL REVERBERATION

Equipment and methods

Top: Typical arrangement of an echo room. If possible the walls should not be parallel to each other. *Middle:* Reverberation plate. The metal sheet is suspended at its four corners inside a wooden box. It is thus reasonably insulated from acoustic sound pickup provided it is not too close to a loud noise. *Bottom:* Reverberation springs. They are usually arranged in pairs, each with two springs of different lengths joined together.

Ambiophony

The use of one of the various artificial reverberation devices described on page 72 can help to produce a tolerable broadcast even from a studio with very dead acoustics but it does nothing to help the performance. Playing an instrument or singing in a dead acoustic can be a depressing business. Performers tend to force their tone and lengthen their notes. If artificial reverberation is added to their sound and they don't hear the result and allow for it, an unnatural effect can result.

Assisted resonance
Surrounding the musicians with reflecting surfaces can help and the use of a good sound reinforcement (PA) system to increase their power and spread it around the room can help even more. Several systems are in use in concert halls (including the Royal Festival Hall, London) that rely on a large number of separate microphone / amplifier / loudspeaker systems each working within a narrow frequency band (the microphones are in helmholtz resonators) to create the effect of multiple reflections.

Ambiophony
The most adaptable, and therefore the most suitable assisted resonance system for television purposes is ambiophony. To produce ambiophony a large number of loudspeakers are arranged around the studio walls and on the ceiling. The output of a special set of microphones placed within about 3m of the orchestra to favour the strings and woodwind is fed to the loudspeakers through a number of time delays. The time delays are arranged to correspond roughly with the time the sound from the orchestra takes to reach the position of each loudspeaker so that they behave rather like natural reflections from the walls. This is achieved by means of a tape loop with one recording and eight reproducing heads. To help diffusion the loudspeakers are not all connected to their appropriate delay but are somewhat intermixed.

Reverberation
A proportion of the output of the longest delay is fed back to the record head to lengthen the reverberation time.

Setting up procedure
The ambiophony system requires very careful setting up and manipulation for every occasion. Once the optimum microphone positions are located the frequencies at which the system can become unstable have to be established and the gain reduced by means of an octave filter (see page 66).

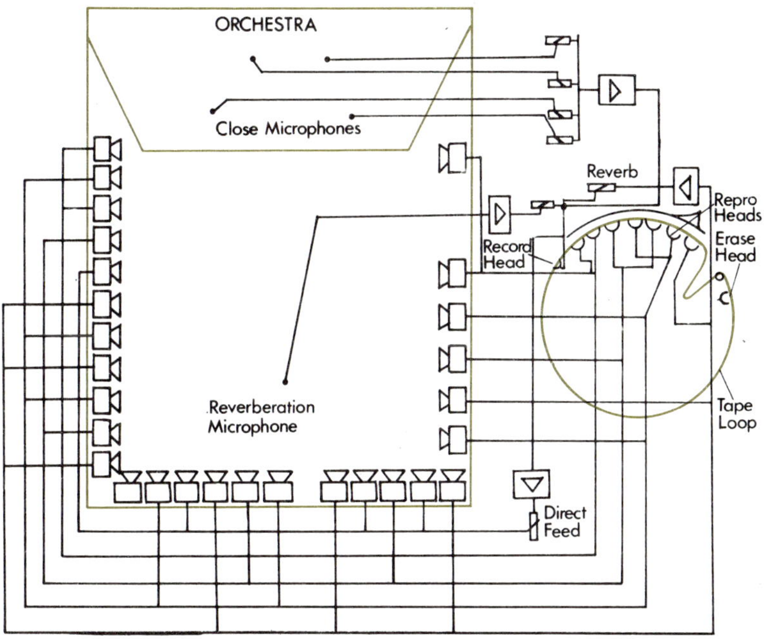

AMBIOPHONY

Installation
The outputs of a number of close sectional microphones are mixed (separately from the broadcast output) and fed to the recording head of a tape loop. The outputs of the eight reproducing heads are combined into four circuits and fed in semi-random manner to a large number of loud-speakers arranged around the walls and ceiling of the concert hall. There is also a direct (undelayed) output to some loudspeakers and an output from a distant microphone (to give a measure of acoustic delay). Some of the output of the last reproducing head is fed back to the record head.

Talks and Lectures

Talks and discussions tend to be among the simpler productions from the sound point of view but that is not to say that they are not important. The essence of a talks programme is the imparting of information and it is essential that the speech is clear.

Introductory Music

Most talks productions begin with introductory music. Unless the talk happens to be about music it is likely that a proportion of the potential audience will appreciate the talk more than the music. If the music is too loud they will switch off or at least turn down their volume controls to the detriment of the rest of the programme. It is important therefore not to allow the introductory music to be too loud. It should normally peak less than the speech.

Microphone arrangements

For a speaker in a static position a slung microphone can be used overhead if a long-shot involving headroom is not required. Alternatively a boom can be used. If the speaker is seated, a floor standing microphone is suitable, perhaps placed to the side just out of shot. If the microphone is likely to come into shot, as in an introductory long-shot, an electrostatic type with one-metre extension tube is less obtrusive.

If the speaker has to walk about at some point and a boom is not available or convenient a lanyard microphone (see page 26) could be used but is not generally recommended because the speech quality is likely to be variable and inferior, owing to the unnatural position of the microphone.

If the speaker is seated at a table or desk a neat microphone may be acceptable in shot. For good speech quality the top of the table should be covered with acoustic absorbing material (e.g. thick felt) to prevent a booming effect caused by reflection from the surface. A still better arrangement is to make the table out of a framework of wood or metal with a fine mesh surface covered with an open-weave fabric so that the sound waves pass through as though it did not exist.

If the speaker is inclined to thump the table the microphone should be suspended from a floor stand and pass through a hole in the table or behind a masking front piece.

1

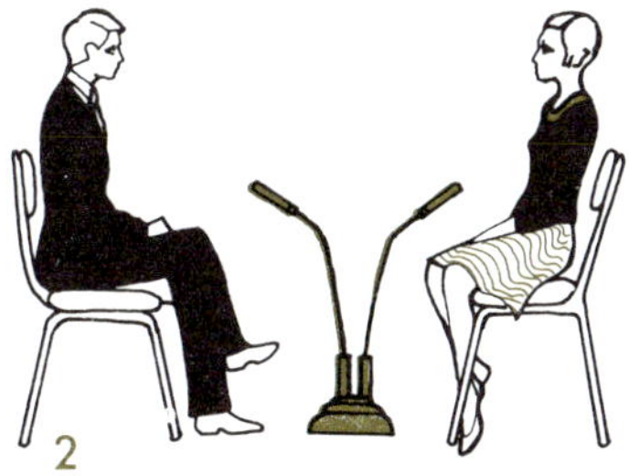
2

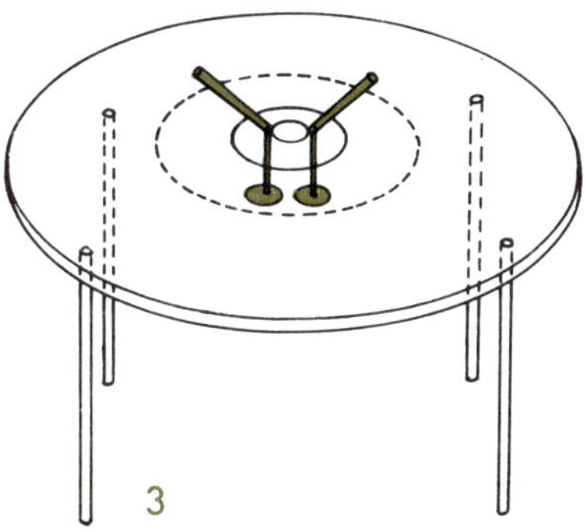
3

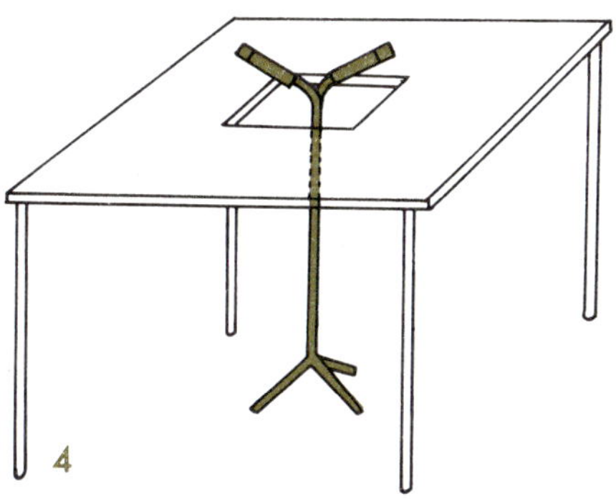
4

MICROPHONE PLACING

Talks and lectures

1, Electrostatic microphone with extension tube suspended over static speaker. 2, A pair of electrostatic microphones with short extension tubes mounted on a double stand for two seated speakers. 3, Circular talks table. The table top is acoustically transparent. A sprung platform is provided below the central hole on which to mount the microphone stand. 4, Table with hole for use with floor-stands.

Informal discussions usually require naturalistic settings and hidden, or at least inconspicuous, microphones.

Informal Discussions

When a number of people take part in an informal discussion it is often required to make the arrangement look like a domestic situation. In these circumstances microphones would look incongruous in shot so the situation usually calls for the use of a boom or booms. However if the discussion is completely unscripted the director may wish to have a wide angle shot available on one camera throughout so that he can cut to it at any time for reaction or because he does not know who is speaking. In this case it may be necessary to employ super-directional microphones, on booms kept above the frame of the long-shot, or to hide microphones in shot. A typical arrangement for an informal discussion would be to have the participants seated in a semicircle around a coffee table. This could accommodate a vase of flowers in which could be hidden several microphones pointing towards the various speakers.

The reason for having more than one microphone in these circumstances is to enable a balance to be obtained between voices of different strengths and also possibly at different distances.

With all these arrangements it is likely that the microphones will be working further away from the speakers than is ideal so it would be advisable to have available a 'presence' filter to restore the presences (see page 66).

Excessive microphone distances are also liable to spoil the effect of an intimate setting by introducing a distant, reverberant acoustic effect. As much sound-absorbing material as possible should be incorporated in the settings. It can also be an advantage to place thick drapes on frames to fill the spaces between the cameras. As well as helping the acoustics, these help the participants by promoting a more intimate atmosphere.

If the speakers are likely to move about or change their direction of voice projection radically it may be necessary to fit them with personal (lanyard) microphones as an additional safeguard.

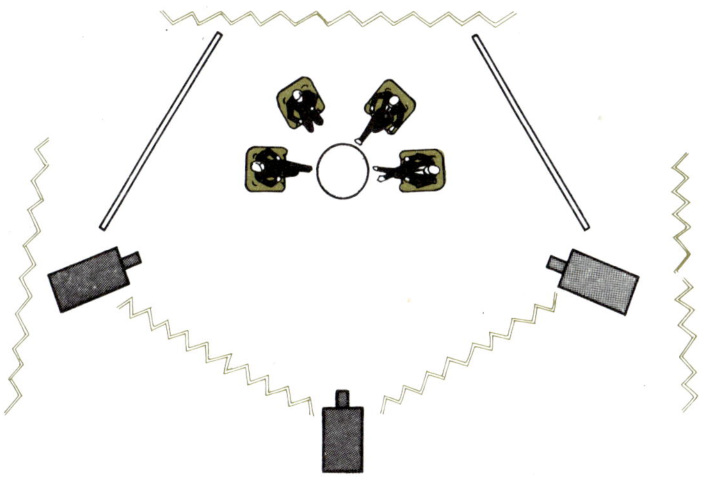

MICROPHONE PLACING

Informal discussions

Top: Vase of flowers hiding microphones. Two directional microphones are used instead of one omnidirectional to assist in balancing between voices of different power. *Bottom:* Set-up for informal discussion in armchairs. The sets are made of absorbent material and thick velour drapes on frames are used whenever possible to insulate the area from the studio acoustic.

Formal Discussions and Panel Games

Formal discussions and panel games tend to adopt rather similar arrangements in television. In either case the participants and the chairman are usually seated at a table or stylised desk. If it is a game (or discussion) in which the people taking part are competing individually (or expressing individual points of view) they are usually arranged to be seated at one desk with the chairman at one end or at another desk facing them. On the other hand, if it is a team game (or discussion), they tend to be divided possibly with separate desks with the chairman seated in the middle. In either case it is generally necessary to give each person a separate microphone in order to balance the levels of the voices and to ensure that each microphone is reasonably close to the person concerned.

This is especially necessary if an audience is involved to guard against PA instability (see page 84) and because it may be necessary to compromise for different angles of projection of the voice. The artists may have to turn one way to face the chairman, another to face the audience, and possibly they will also want to turn to talk to each other. If considerable changes of angle are involved it may be necessary to employ extra microphones at the extreme ends of the desk or even, as a last resort, to use personal (lavalier) microphones.

There is usually no objection to showing microphones in shot for this type of programme. Small desk stands can be useful, but there is always a risk of people banging the desk and causing noise by conduction — a risk that is multiplied by the number of persons sharing the desk.

This problem can be overcome by using floor stands arranged so that the microphones enter through holes in the desk without touching it or stand in front of the desk. A masking screen hides the stands in longshot views.

MICROPHONE PLACING

Formal discussions and panel games
1, Typical grouping for two opposing panels and chairman. Note angling of microphone to take account of artist's projection. 2, Panel game with audience. Note requirement for extra microphones to take account of artist's alternative direction (to chairman or audience). 3, Section view of panel desk showing microphone stand screen.

Planning for Studio Audiences

Many Variety and Situation comedy productions involve a studio audience. Notwithstanding the fact that most television studio audiences are admitted free they should be treated with the utmost respect and their presence taken into careful consideration at the planning stage. A happy audience is not only a good public relations exercise it is a vital complement to the production.

The audience should be arranged in as compact a formation as possible. This is to assist with PA and audience reaction coverage (see page 88) and because they tend to stimulate each other better in close formation.

Visibility

The action should take place as close to the audience as possible. This is usually quite easy to arrange for variety but difficult for situation comedy if a large number of sets is involved. It is useless to arrange the action all around the studio walls remote from the audience and then expect them to react. Even if the action is reasonably close to the audience great care must be taken not to mask peoples' view more than absolutely necessary. Booms tend to be the worst offenders from the masking point of view and when these are used they should, whenever possible, work from the sides of the set.

Operator discipline

All operators involved in audience shows should avoid distracting the audience's attention when they are not immediately involved, e.g. preparing for the next sequence. The audience can easily find the 'mechanics' of the operation more fascinating than the performance. All 'off stage' movements during a performance must be as unobtrusive as possible and preferably screened from view.

Production routine

The audience should be ushered in to subdued background music (non vocal) and a microphone made available to the producer before the show for 'warm up' and afterwards to thank them and send them away happy.

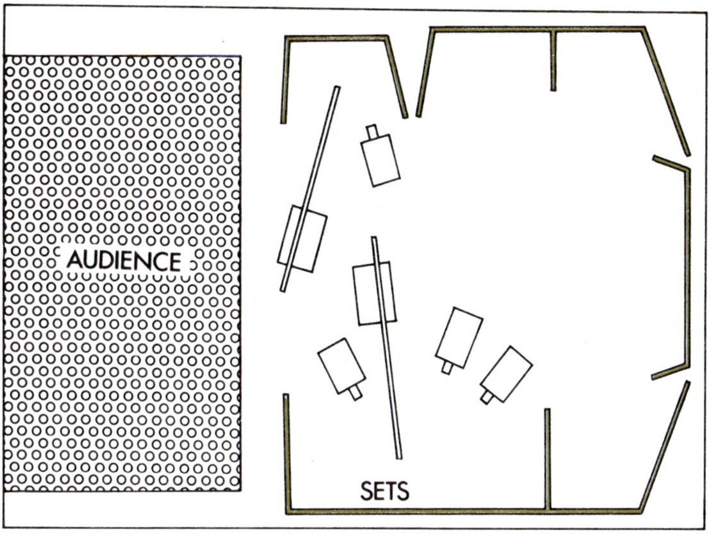

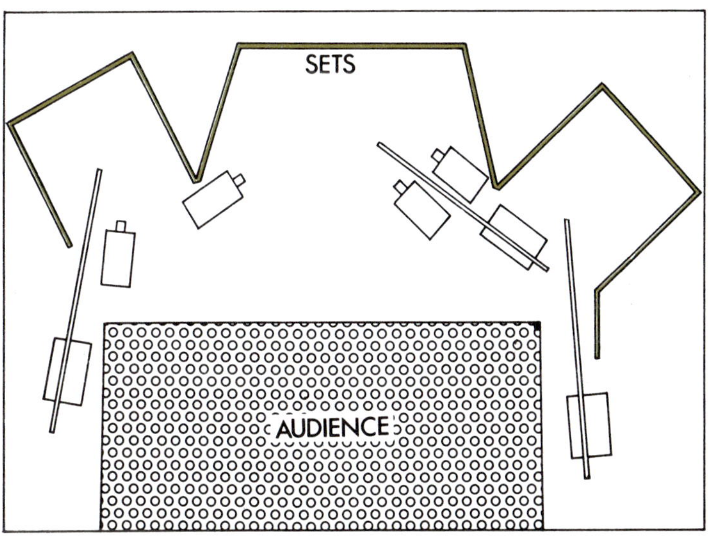

STUDIO AUDIENCIES

Set arrangement
Top: wrong arrangement of sets for a situation comedy programme. The action is mainly screened from the audience who get a distracting view of the technicians. *Bottom:* Correct arrangement for programme audience. The cameras and booms work from the side as much as possible.

Audience sound reinforcement is vital to the performance of light entertainment productions.

Audience Reinforcement

One of the most essential aspects of any production involving an audience is an efficient system of audience reinforcement or, as it is usually called, public address. There are two basic types of system:

High level system

For high-level operation a small number of directional loudspeakers (sometimes only one) operate at high level to supply an audience at some distance. The main disadvantage of the high level system is that feeding a high level of sound into the auditorium tends to excite the reverberation and to turn it into a vast echo room. Also, although directional loudspeakers are used, any small reflecting surface in the vicinity can spoil the directivity and at high level this might cause feedback to the microphone resulting in a howlround.

Low level system

For low-level operation, a very large number of small loudspeakers operate at comparatively low level to supply a small section of the audience. These are usually simple small box baffle loudspeakers. Some systems even provide a loudspeaker fitted in the back of each chair. The most common type of directional loudspeaker used for PA systems is the column or line-source. This consists of a number of loudspeakers arranged in line in an acoustic baffle and connected to operate in phase. The back of the baffle box is usually either completely closed or provided with an acoustic resistance. This causes the sound to take approximately the same time to emerge from the back as to travel around from the front so that the two tend to cancel at the back.

The line source loudspeaker tends to have a circular polar diagram at right angles to the width of the column and a narrower lobe along its length.

Line source loudspeakers tend to be more directional in the higher frequencies. This can be counteracted by various means such as curving the column, tapering the input to decrease towards the ends by means of a transformer or using more closely spaced treble loudspeakers.

Compromise arrangement

In television where microphone distances can be excessive, the PA problem is often critical and can usually best be met by a compromise arrangement of directional loudspeakers used in large numbers almost in the low level manner. In television studios the best method is usually to employ four-foot column loudspeakers mounted in stirrups that can be conveniently suspended over the auditorium on special hoists or on spare lighting barrels.

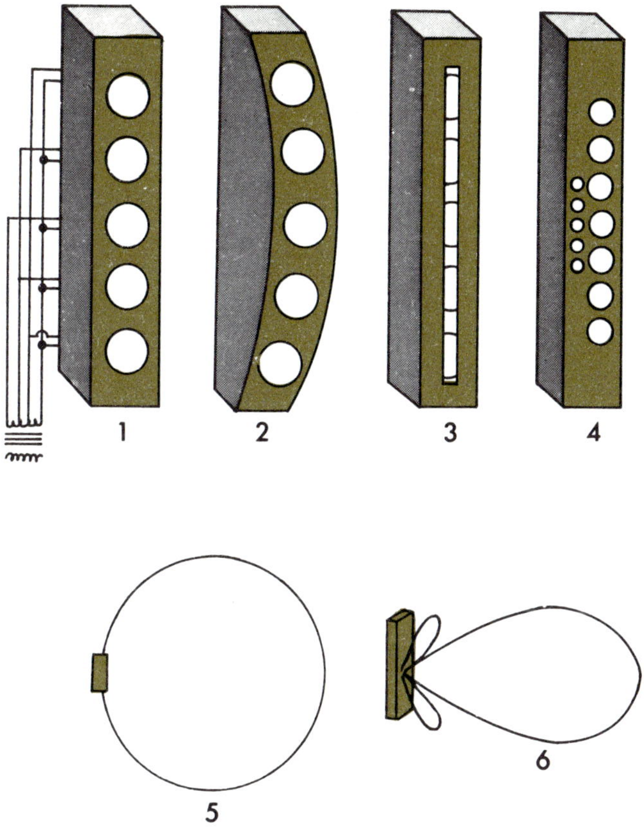

AUDIENCE REINFORCEMENT

Line source loudspeakers
Methods of balancing the treble and bass directional characteristics: 1, Tapering the input to favour the middle. 2, Curving the face of the box. 3, Causing the loudspeakers to speak through a slit (this widens the HF polar responses). 4, Employing closer-spaced HF units (less directional effect). 5, Polar response of column loudspeaker in horizontal plane. 6, Polar response of column loudspeaker in vertical plane.

PA Connection Points

The need for an efficient system of PA and suitable loudspeaker arrangement were discussed on page 84. There remains the matter of how to connect it to the output of the broadcast microphones.

Before the fader PA
In modern practice programme control is exercised by the individual channel faders. The PA is therefore arranged to be connected at full strength immediately the faders are moved off the backstop.

The main advantage of this system is that the PA is not subject to programme control so that the loop gain of the potential feedback system (microphone — amplifiers — loudspeakers — microphones) remains fixed and unless the position of the microphone is altered it should be possible to work close to the howlround point without undue risk. Also the fact that the artist is able to hear his volume increase as he approaches the microphone can encourage him not to stray too far from it. The disadvantages are that the sound on the PA may not match the broadcast sound so that any pick-up on the broadcast microphones may be detrimental. Also any sound that is established before the fade-up will come in with a sudden burst.

After the fader PA
If the PA is connected after the fader, it is subject to programme control unless this is exercised only by the group fader. This can present an increased risk of howlround or variable colouration. It can also result in a rather disturbing effect for the audience if the nearest loudspeaker is not in direct line with the artist. The sound can appear to hop about between the two as the relative volume varies.

Combined arrangement
A method that can combine the best features of before and after the fader PA is to divide the control into two sections and feed the PA from the connection between the two. In this way it becomes subject to the lower half of the control function and therefore is faded up gradually but is not affected by subsequent programme control.

The ideal arrangement provides PA connection that can be switched between mid-fader, after the fader, and after the group fader.

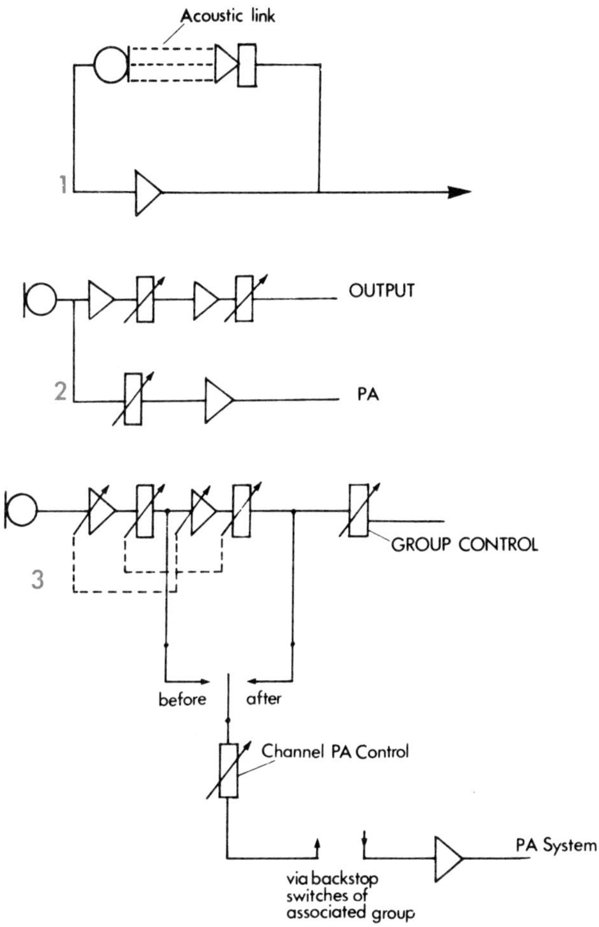

PUBLIC ADDRESS SYSTEMS

Points of connection
1, A public address system is intrinsically a feedback arrangement. The liability to instability depends on the loop gain of the amplifiers overcoming the acoustic path loss. Obviously it is possible to work nearer to the 'howl-round' point if both factors are fixed. 2, Simple before-the-fader arrangement. 3, Sophisticated arrangement providing a choice of before or after the fader PA.

Audience reaction encourages the performer and involves the viewer.

Audience Reaction

Audiences are, in general, invited to television shows for two purposes:
1. To encourage the artists and assist them in judging the effectiveness and timing of their material.
2. To provide laughter and applause for the broadcast. The first requirement therefore is that they can see the artist clearly (see page 82) and the second that they can be clearly heard.

Sound character
The type of sound usually aimed at for audience reaction involves two characteristics that tend to be conflicting:
1. The audience should sound large and impersonal (i.e. no individual should be prominent — especially if he has a distinctive laugh)
2. The audience should sound reasonably close and intimate in character. The ideal being to give the listener the impression that he is seated in the middle of the audience. This effect can usually only be obtained by using a large number of microphones each one at a distance of only two or three metres from the particular group of audience it is intended to serve.

Separation from PA
A major problem in designing an audience reaction microphone set-up is minimising pickup from the PA loudspeakers which are serving the same area. Sound picked up in this way adds colouration from the uneven response of the PA loudspeaker to the broadcast sound and makes it sound reverberant, turning the auditorium into a sort of large echo room. This can best be avoided by using directional loudspeakers and microphones with their dead sides facing each other.

'Milking' the audience
It is unlikely that the desired effect of audience reaction will be obtained for most comedy production. Applause is usually much louder than laughter but should be broadcast at a rather lower volume. Some weak laughs need to be magnified somewhat.

It is therefore often necessary to 'milk' the audience. That is to 'swell' the volume of laughs by bringing up the volume after the laugh has started, increasing the natural rise and fall, as realistically as possible.

It is important to reduce the level before the artist speaks again otherwise his voice will be picked up by the reaction microphones producing an echo effect.

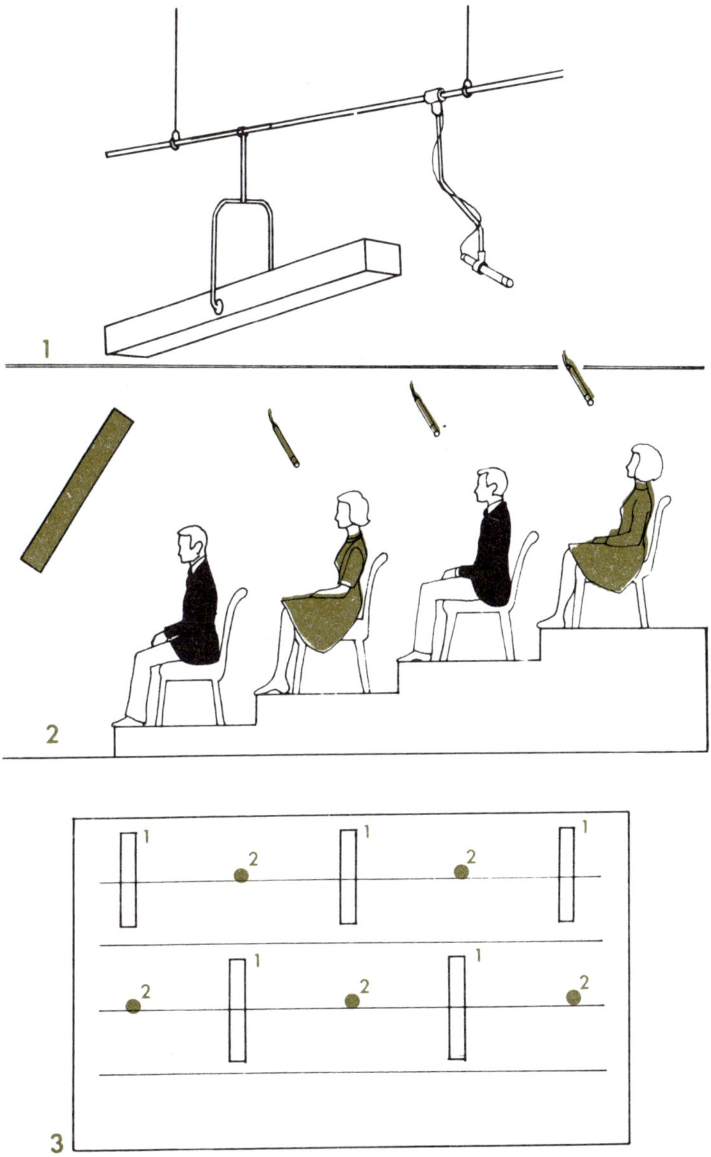

AUDIENCE REACTION

Loudspeaker and microphone arrangement

1, Arrangement of PA loudspeaker and audience reaction microphone on lighting barrel. 2, Microphone arrangement for a more distant loudspeaker (high-level) system. 3, Plan view of audience area showing disposition of loudspeakers and microphones.

Audience Participation

Some of the most difficult types of discussion programmes from the sound point of view are those where members of the audience are invited to take part. It is quite easy to arrange for a camera to zoom in from a wide angle shot of a large audience to show a close-up of a single speaker but picking up the corresponding sound can be a very different matter.

Set up arrangement
The simplest set-up occurs when it is possible to arrange that only a few specified members are allowed to speak. These can then be placed in convenient positions where they can be covered by suitably suspended or stand microphones.

Hand held microphones
In circumstances where any member is able to speak, but there is good notice of the order, it may be possible to pass around a hand held microphone. A number should be made available each to cope with a small block of seating.

Rifle microphones
If the discussion is *ad lib* but not very quick-fire, coverage can often be achieved by planting sound operators with hand held rifle microphones among the audience each to cover a few rows of seats.

Booms
The best arrangement for a free-flowing discussion is usually to use booms equipped with rifle microphones. The reason for this type of microphone is that the booms are unlikely to be able to reach all members of the audience and will probably have to work at a hight sufficient to allow the cameras to be able to take wide-angle shots without seeing too much of the booms or their shadows. Several booms will be needed to cover a large area. This also enables one boom to search for the next likely participant while another is covering the current speaker.

Parabolic microphones
Where the microphone coverage has to be inconspicuous and provided that the auditorium is fairly short and steeply raked, good results have been obtained using a combination of microphones with parabolic reflectors and slung conventional microphones. The slung microphones provide a basic "blanket" coverage (as for audience reaction) albeit too far from any individual, and the "presence" is added by mixing in a suitable proportion of a well aimed parabola microphone from which the bass has been rolled off. In this way it is possible to make the sound appear to zoom in with the camera.

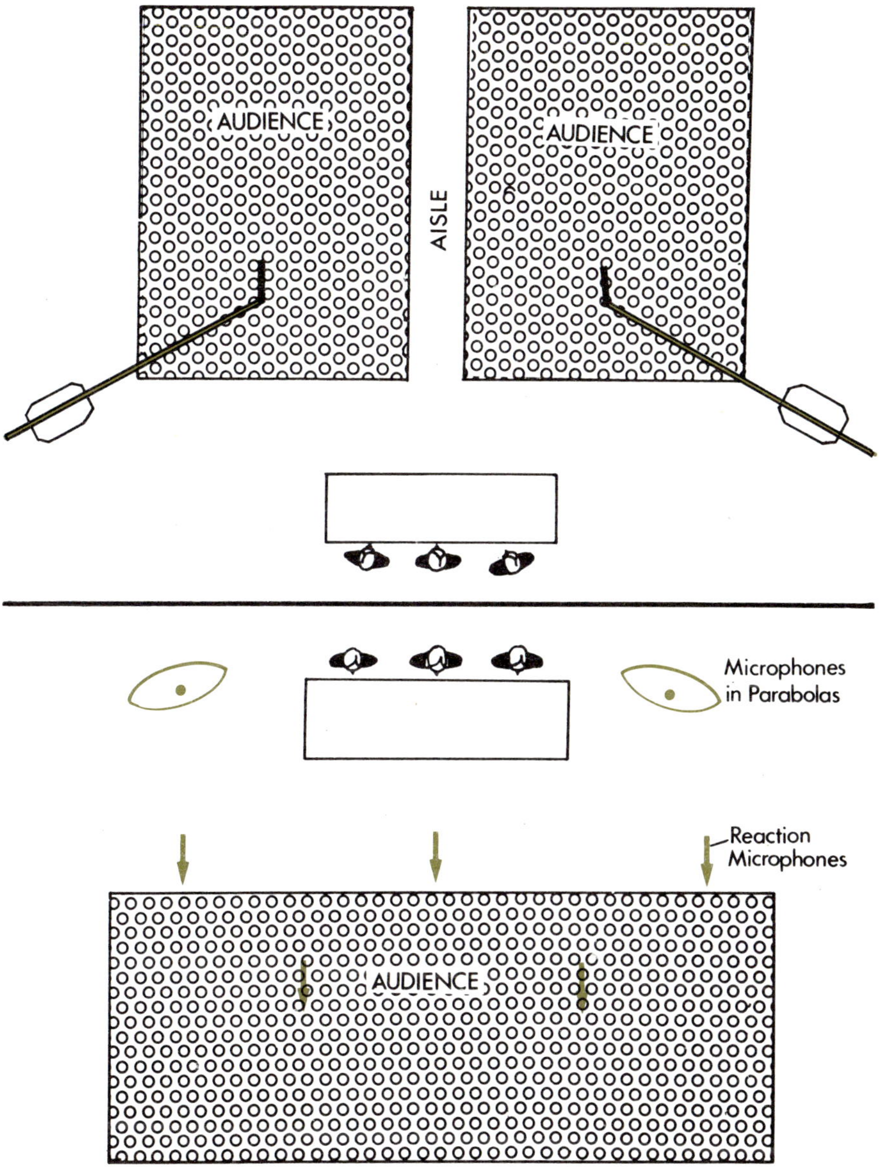

AUDIENCE PARTICIPATION

Methods of coverage
Top: Audience grouped into two compact areas each covered by a boom equipped with a 'rifle' microphone. *Bottom:* Audience coverage by combination of reaction microphone and parabolic reflectors.

Discussions between Remote Centres

In television it is quite often required to link up people in different areas in such a way that each can see, or at least hear the other and the viewer can see and hear both.

Deaf aids

The major problem in achieving two way communication using loudspeakers at each end is to prevent so much pickup of the loudspeakers on the microphones that the sound will have a hollow coloured quality. With loudspeakers at both ends it is possible even for the whole loop to go into self oscillation with disastrous results. The simplest way of overcoming this is to provide the participants with deaf-aid earpieces (being less obtrusive than earphones).

Loudspeakers

If loudspeakers are used the following rules should be adopted:

1. The loudspeakers should be fed from a 'clean feed' source. This supply, as its American term 'mixed-minus' more graphically suggests, provides each participant (and there could be a number of sources) with everybody's sound output except his own.
2. Directional microphones, and if possible directional loudspeakers, should be provided each with its directivity pattern exploited to obtain maximum separation from the other.

In practice it has been found that in reasonably dead acoustics the best results are achieved by using a cardioid microphone as close to the artist as possible with a loudspeaker at a distance of at least two metres pointing toward its dead side.

3. The settings surrounding the artist must be as sound absorbing as possible and not so angled that they reflect the loudspeaker sound back on the live face of the microphone.

Line up routine

When several sources are involved, each using loudspeakers, it is essential to adopt a strict routine in which the loudspeaker levels are set for each for just comfortable listening before their respective microphones are faded up. Any subsequent re-adjustment could cause one location to 'steal' more than his share of the available volume and cause the others to reduce their gain to avoid a howlround.

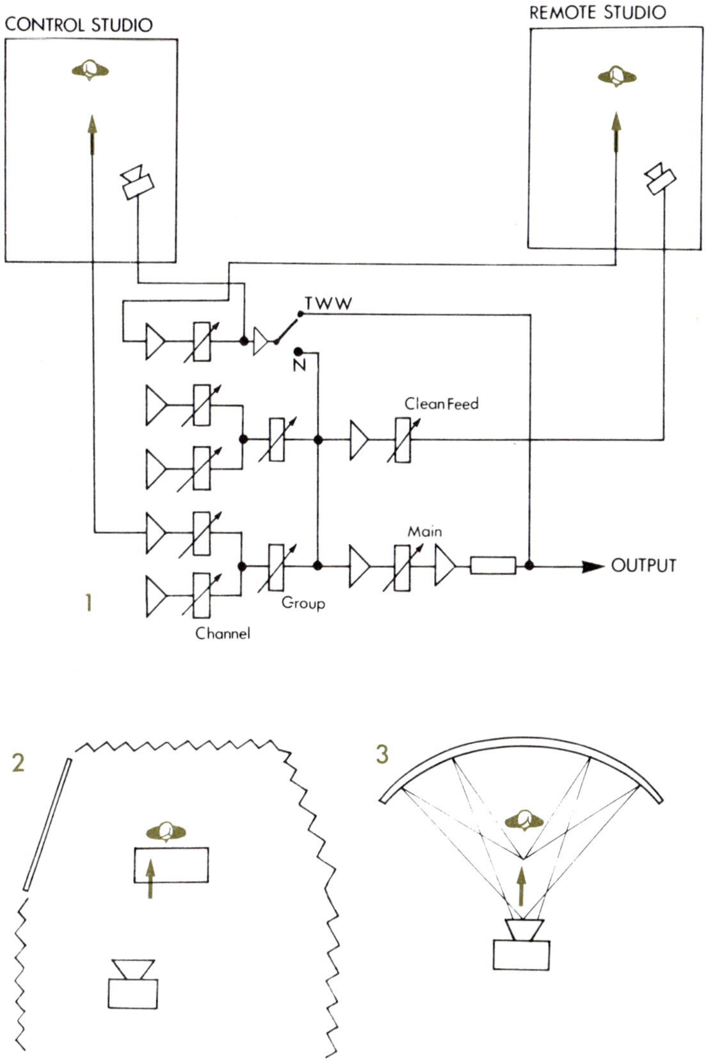

Alternative arrangements
1, Arrangement for deriving 'clean feed' (American term 'mixed-minus') for two-way working with a remote source. 2, Sensible arrangement for remote interview studio. The loudspeaker is at some distance from the dead side of the microphones. The settings are as absorbent as possible. 3, Bad arrangement: the concave reflective settings focus the loudspeaker output on the live face of the microphone.

Planning for Drama

Planning the sound coverage of a drama production should begin many weeks in advance of the studio date by a careful reading of the script and discussion with the director to find out
1. The character of the story, the period, pace and general degree of sophistication and complexity to be employed in the production.
2. The requirement for filming or electronic pre-recording, possibly of exterior sequences, and the need for matching with the studio sequences.
3. The requirement for incidental music and recorded sound effects and whether they are already available or will need to be specially recorded in advance.
4. The need for special effects or 'props' such as 'practical' (working) telephones or intercoms etc.

Filming or pre-recording
If filmed or pre-recorded sequences are involved and required to match the studio sound it is advisable to find out which medium is likely to be faced with the least flexible situation and then arrange for the other to adopt a similar technique. For example it is no use the studio taking advantage of the opportunity to use very close microphone technique if the pre-recorded sequences with which it is to be intercut are forced to work at a distance. A compromise should be adopted to make the transition less obvious.

Planning meeting
The sound supervisor should attend at least one full-scale planning meeting several weeks in advance of the production. This would include the director, the lighting director, scenic designer, wardrobe and make-up.

The sound supervisor should check the arrangements of the sets, the camera positions and, in consultation with the lighting director who can give an indication of the direction of the key lights, decide on the best method of obtaining sound coverage.

This is achieved by reference to the plans, elevations and models as outlined on page 50. The space occupied by the boom and its coverage can be ascertained by the use of a Perspex cut-out. It may be necessary to arrange the juxtaposition of the various sets to assist in boom movement or to provide sufficient acoustic separation for perspective effects etc. This is usually achieved by cutting out the outline of the sets on tracing paper and moving the pieces around on the plan. When the position of booms and other microphones is decided they are marked on the camera plan.

Once the sound coverage has been established the sound supervisor can mark up his script to show the sequence of sources.

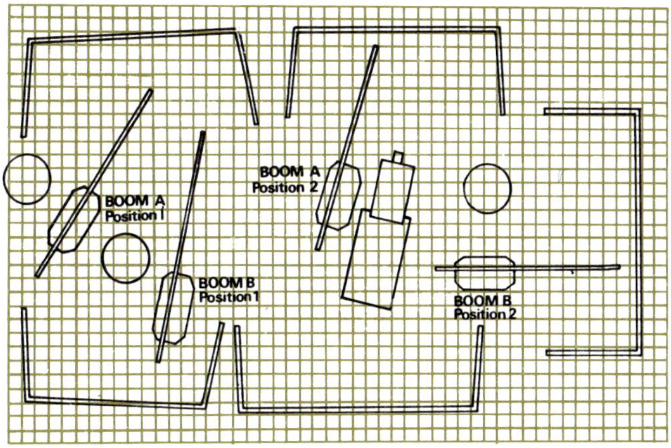

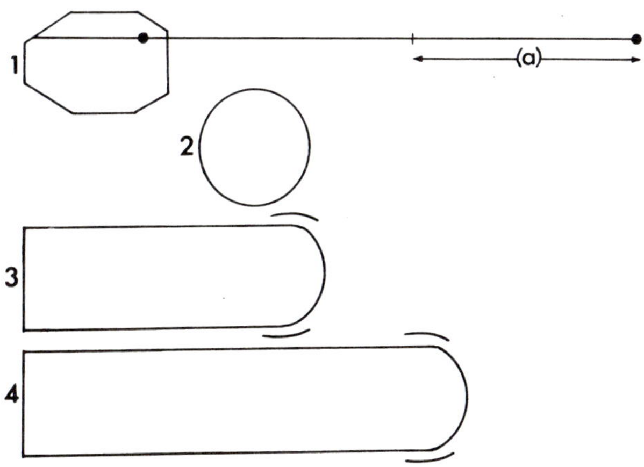

PLANNING FOR DRAMA

Floor plans and cutouts
Stencil cutouts of production equipment are used for planning in association with studio floor plans, to the scale of ¼ inch to the foot (or 6 mm to 30 cm). 1, Large boom: (a) range of extension 10-20 ft (3-6 metres). 2, Vinten camera pedestal. 3, Vinten Heron camera crane. 4, MPRC camera crane.

Co-ordinating Sound and Picture

In drama, more than any other type of production there is an attempt to achieve a realistic effect. It is important to remember that sound can set the scene just as well as, and sometimes better than, vision. Listeners to radio drama, or anyone who has tried the simple expedient of listening to natural sounds with his eyes shut are well aware of the ability of sound to conjure up atmosphere and a sense of environment without need of a picture. This must be equally true of television sound, whether the viewer realises it nor not. If the scene set by the sound and the picture do not agree, the viewer tends to follow the picture but is left with an uneasy sense that something is wrong. On the other hand, if the two media match perfectly they combine to produce a satisfying sense of realism that has more impact even than the sum of the two senses.

Perspective

An important property of sound is its ability to suggest perspective, a sense of distance or third dimension, and this also must be in agreement with the visual perspective shown on the screen. Few things are as destructive to dramatic realism as to hear an artist appear to recede in sound as he approaches in vision, or vice versa. Such an effect can easily be produced accidentally e.g. if an actor, approaching the camera, overruns the minimum retraction of the boom.

Distance

The ear tends to judge the distance of a sound source in the following ways:

1. By relative volume level. The ear has a poor memory for absolute level but is quick to notice a change. Obviously the louder the nearer.
2. Tonal characteristics. With the voice the sibilants and higher consonants are more pronounced at close range (hence the so- called 'presence filters' which increase output in the upper-middle frequency range).
3. Acoustic effect. More reverberation is associated with greater distance.
4. Frequency response. We recognise that in most real-life situations high frequencies are more affected by distance than low.
5. Time factor. The ear can detect the extent to which the direct sound precedes the reverberation and thus distinguish between a close sound in a reverberant room and a distant sound in a less reverberant one.

Perspective matching can be achieved by careful control of studio and set acoustics, microphone positioning, the addition of artificial reverberation and the subtle use of frequency correction.

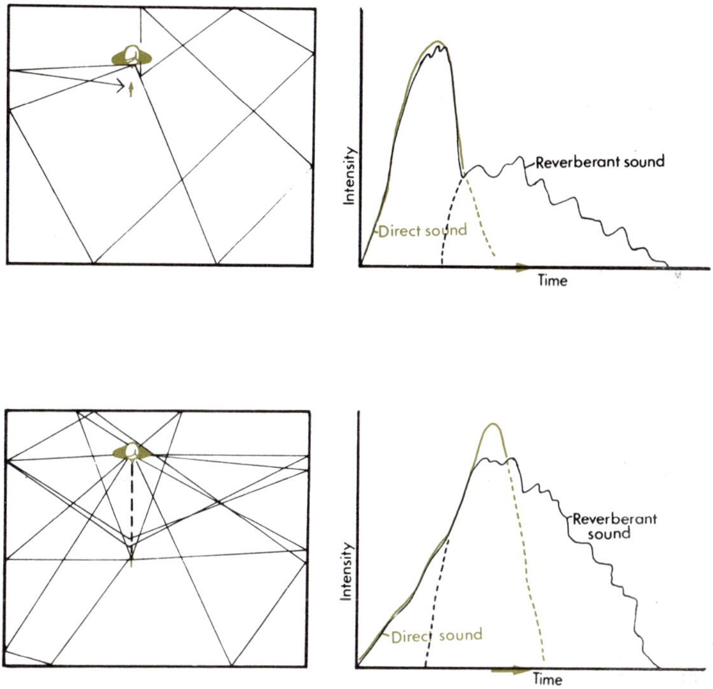

AURAL AND VISUAL PERSPECTIVE

Distance and reverberation
Top left: Artist working close to microphone. His direct sound reaches the microphone well in advance of the reverberation which is at much lower volume. *Top right:* The sort of decay curve to be expected from an artist making an impulsive sound at close range in a reverberant room. Our ears can recognise that the direct precedes the indirect sound and this gives us the clue that the sound is close, but the room is reverberant. *Bottom left:* Artist at some distance from the microphone. The reflections arrive at a similar time to the direct sound, which appears to be submerged in the reverberation. *Bottom right:* Slower build-up and later decay of sound. Our ears associate this effect with distance.

Dialogue and Split Sound

Most drama involves dialogue i.e. people, talking to each other. If the coverage is by means of a boom it is usually possible to cover a number of people with one microphone by splitting the difference between them, provided that:

1. They are not much more than about two metres apart.
2. The shots of the individual actors are not extremely close and the dialogue intimate.
3. The people are reasonably equidistant from the cameras and of fairly similar height.

When these conditions are met the usual position for the boom is above and downstage, i.e. toward the cameras but splitting the angle between them unless it is necessary to favour one of the artists to compensate for a weak voice or match an exceptionally close shot.

Split sound

Problems can arise when the artists are separated by distances much greater than two metres. If they merely go towards opposite sides of the set and are pictured entirely in long shot it still might be possible to cover both with a boom held high between the two. However, as soon as a camera takes a close-up shot of one of them it will be necessary to bring the boom in close to cover him. If another close-up shot follows quickly it is most unlikely that the boom will be able to swing over in time. It will therefore be necessary to 'split the sound' by bringing in another boom so that each can cover a section of the area in turn.

Up stage — down stage split

While it is usually quite simple to use two booms on one set with a sideways split it may not be so easy when the artists split between front and back of a deep set so that one is seen over the shoulder of the other. In this case there is considerable risk of seeing the upstage boom in the shot of the downstage artist. It can be useful to have available another method of picking up the sound close to the up-stage artist e.g. a personal or hidden microphone which can be mixed in with the output of his (too high) boom to restore presence in the close-up.

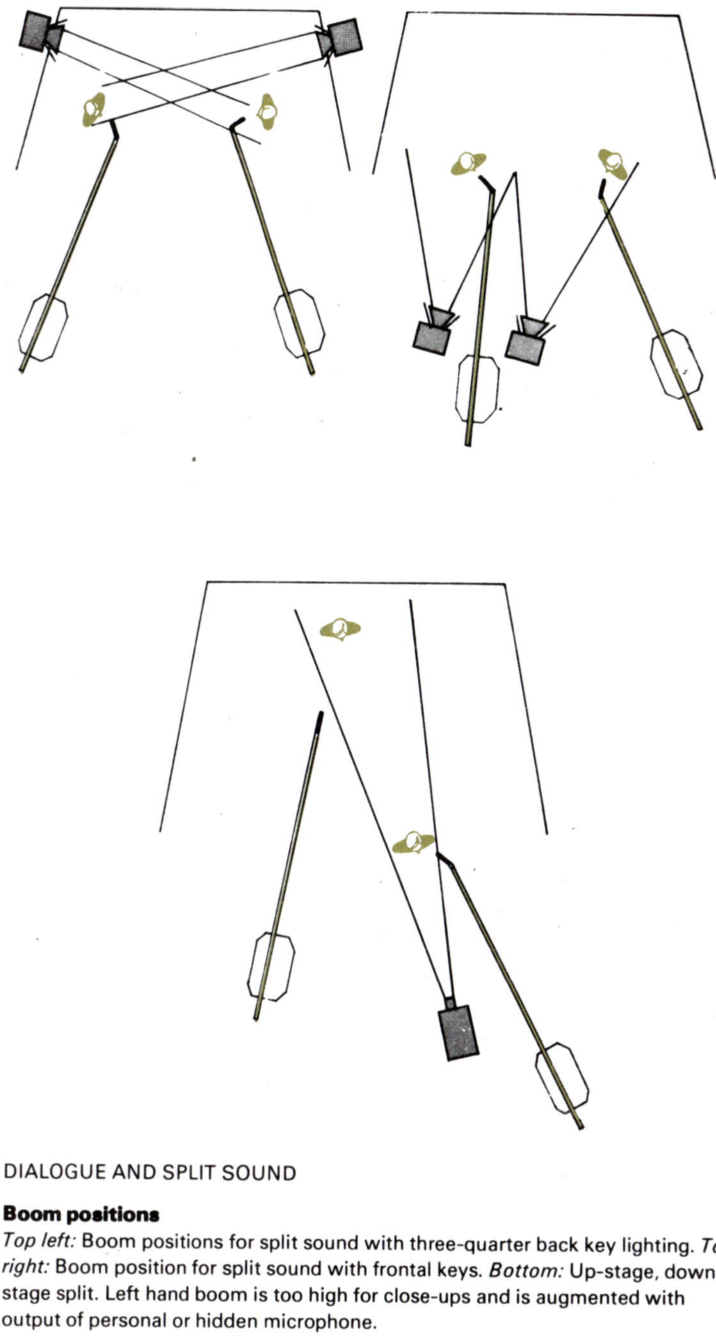

DIALOGUE AND SPLIT SOUND

Boom positions

Top left: Boom positions for split sound with three-quarter back key lighting. *Top right:* Boom position for split sound with frontal keys. *Bottom:* Up-stage, down-stage split. Left hand boom is too high for close-ups and is augmented with output of personal or hidden microphone.

Grouping Problems

When crowds or large groups are involved in drama productions they are usually composed of extras with no lines to say except the proverbial "rhubarb-rhubarb" background chatter. This can produce a very unrealistic effect unless they are told to speak natural conversation at a natural level of voice. The difficulty then usually arises that the actors, who may be close to the crowd, or even mixed up with it, have to be clearly heard above the crowd and moreover no irrelevant dialogue from members of the crowd must be sufficiently distinct to be distracting. The best way to achieve a natural effect in these circumstances is to make a recording of the crowd using natural voices but with such an overall balance that the words tend to merge together. This recording is played back to the crowd during the action and they are instructed to go through the motions of speaking with very little sound so that their live voices merely augment the recording. The sound supervisor then has complete control and can arrange for the actors' dialogue to overlay the realistic crowd noise with sufficient clarity. It is usually advisable to employ two booms, one to remain fairly high for an overall atmosphere effect (the extras movement noises may require synchronous sound more than their blended voices) and another to drop in close to the actors when in close-up.

Tables
Quite often the action in television drama centres around a table. It is usual to arrange the actors, or at least those who speak, around three sides to allow access for the cameras. This usually results in a deep but close form of grouping in which it is very difficult to obtain close-up sound to match close-up pictures. The problem is to avoid shadows (from the close cross-key lighting) and the necessity for the boom to rack and turn quickly. It is usually advisable to employ two booms, each looking after a defined area. If long shots are involved it may be possible to hide a slung microphone in a lamp fitting over the upstage end of the table.

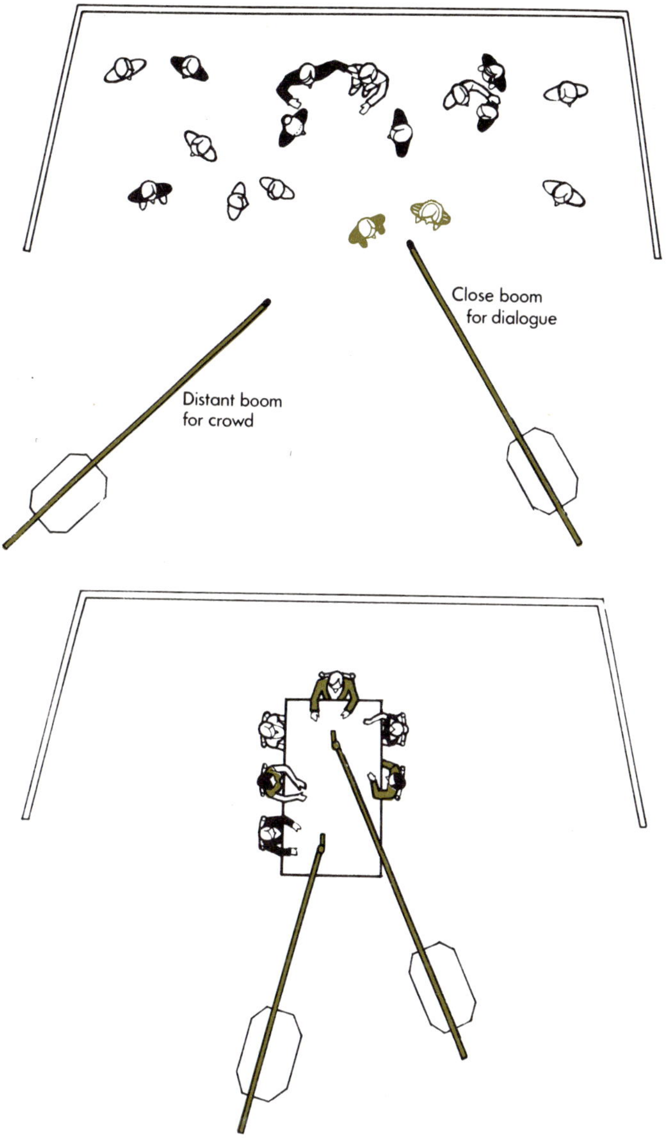

GROUPING PROBLEMS

Using two booms

Top: Crowd scene. Two booms are used, one close-in for dialogue and one high for general crowd atmosphere noises. *Bottom:* A deep arrangement around a table. Two booms may be required but the 'upstage' one has to work at some height to avoid throwing shadows on the downstage actors.

Problems of differing heights between artists can often be overcome by working the microphone to the side of the frame.

Height Discrepancy

A problem can arise when there is dialogue between somebody standing up and somebody seated, especially if the standing person has the stronger voice. If the shots are not too close it may be possible for the boom to work from a position well to the front to take advantage of the cone angle of the camera lens or to work to the side of the frame nearest the seated person.

Long shot to upper room or balcony
There are occasions when dialogue occurs between someone at ground level and a person upstairs, perhaps on a balcony, staircase, or upstairs room. Often it is required to see the higher person over the shoulder of the lower one. This is likely to be a split sound situation the upstage actor taken on a boom tilted high (the maximum height of a boom microphone with the arm at full rack is about 5 metres) or hidden in some convenient set dressing such as a lamp fitting or balustrade. It may be difficult for the other boom to approach the downstage actor sufficiently without coming into shot in which case it might be advisable to work to the side of the frame or use a fishing rod or rifle microphone held underneath the shot.

Action on rostra
When the action takes place above studio floor level e.g. when low angle shots are called for, or the action has to be continuous to an upstairs location, it may be necessary for some of the cameras to work on the rostra also. If there is not room for the boom to do so as well it may be possible to fix a spigot to the scaffolding supporting the camera platform on which the boom arm can be mounted. This is rather a restricting situation and can be an unnatural operating position, so on most occasions a fishing rod would be preferable.

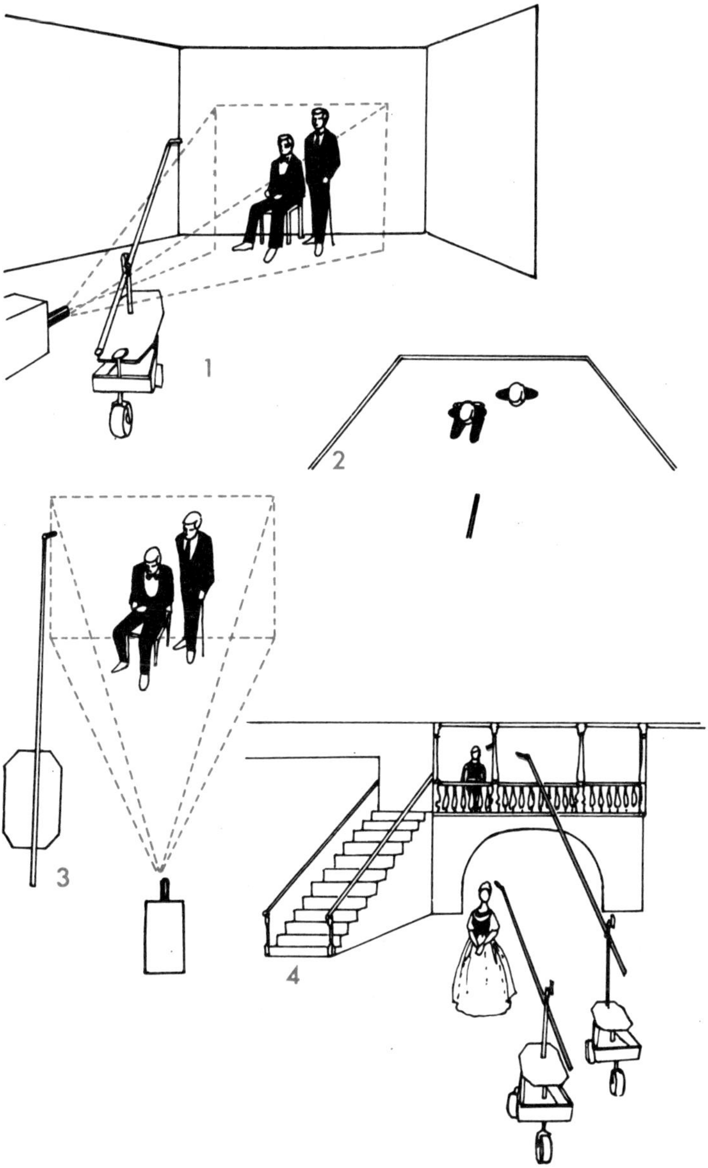

BOOM PLACING

Height discrepancy

1, Boom operating well forward to take advantage of low-angle shot. 2, Plan view showing approximate position of microphone. 3, Boom working to side of frame. 4, Close-up and long-shot coverage for high split-sound situation. A microphone behind the balustrade can be added for 'presence' when the upstairs actor is in close-up.

Ceilings, Beams and Windows

It is sometimes desirable to provide sets with a ceiling, or at least a partial ceiling to obtain a more realistic long shot or allow for a low angle upward-looking shot. This can present a considerable problem for lighting, owing to the difficulty of back-lighting the artists (see page 44), and for sound. One way of overcoming it is to make the ceiling from a succession of coffers or beams which, when viewed from the front, look like a continuous ceiling but which allow space between to shine backlight and possibly to position fixed microphones to cover the action underneath. Alternatively a thin gauze can be used stretched over the area. If it is light in colour and lit from the front it will appear to be solid but a boom or other microphone can be used behind it very effectively. It is, however, difficult to shine backlight through a gauze without spoiling the effect and exposing the position of the lamp. Where low beams are included in the set it may be possible for a boom to work effectively over the top of them, where possible working parallel, between the gaps.

Windows

Windows in television productions usually consist of empty frames without glass. Obviously the frame has no effect on the sound picked up by the microphone, but, if actors are to be seen speaking their lines apparently through a closed window it is necessary to preserve the illusion of glass by modifying the sound. This can be achieved most effectively by dropping the sound level by about half the normal and providing high frequency attenuation of about 3dB per octave whenever the shot is supposed to be through the glass.

Reverse angles

Quite often windows at the back of the set are used to provide a reverse angle viewpoint for the camera. In this case it is necessary to provide a microphone outside the window, in line with the camera shot and fade to it when that camera is used, to prevent a reversal of perspective between picture and sound.

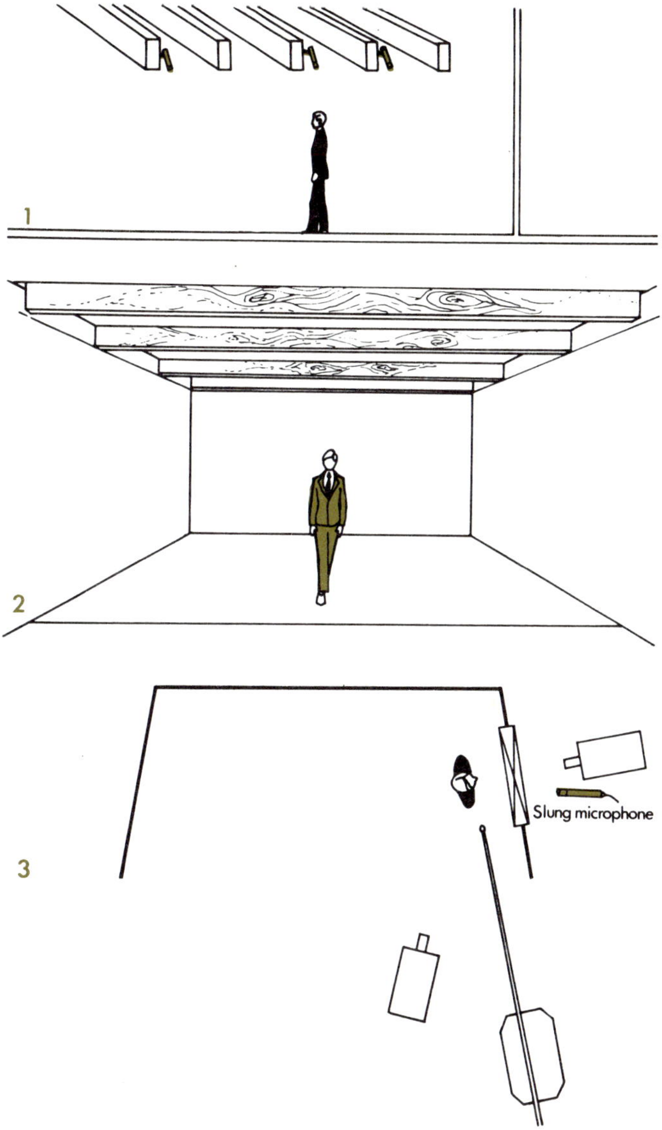

CEILINGS AND WINDOWS

Microphone placing
1, Side view of set showing suspended beams, some of which hide microphones. 2, From the front the beams take on solid appearance. 3, Reverse angle shot through window showing the need for a microphone in line with the camera.

Columns, Arches and Doorways

Columns are sometimes included in television design for architectural reasons or merely to give a sense of depth to the set. They may be supporting scenery or standing on their own, possibly steadied by wire from above. They can cause considerable problems for the boom, particularly if the artists walk past them while speaking. If the artist walks close to the columns, whenever possible the boom should work from the same side as the artist. Even then there is a considerable risk of microphone shadows being cast on the columns, as the lighting will probably also be from that side, so that the boom may have to work at some distance in front of the artist. If the boom has to work from the side of the column it will be very difficult to prevent a break in sound continuity as the boom has to rack in, pass each column in turn, and rack out again. In these circumstances it should either be arranged that the artist does not speak as he passes the column or two booms must be used, one taking over from the other.

Arches

A similar situation occurs when an artist has to walk through an arch while speaking. If the shot remains of constant length from the front so that the artist approaches the camera it may be possible to cope with one boom on the camera side of the arch, but if the camera tracks through the arch, or zooms in to close-up on the other side, two booms will be necessary to cover the walk, one either side of the arch. In this type of situation it is a great help if the sound supervisor can see into the studio so that he can judge the best point to make the change-over. If he is not in a position to see into the studio the boom operator should warn him over the reverse talkback as soon as he has handed over to the other boom and can be released.

Doorways

Doorways present a similar situation to arches except that there is more often the possibility of a reverse angle shot from the other side of the door. This is usually a convenient place for a fixed-position slung microphone.

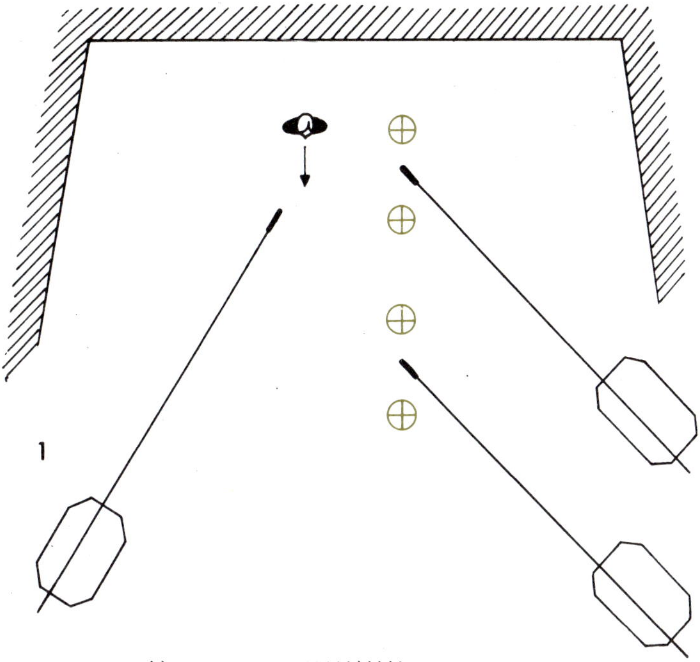

1

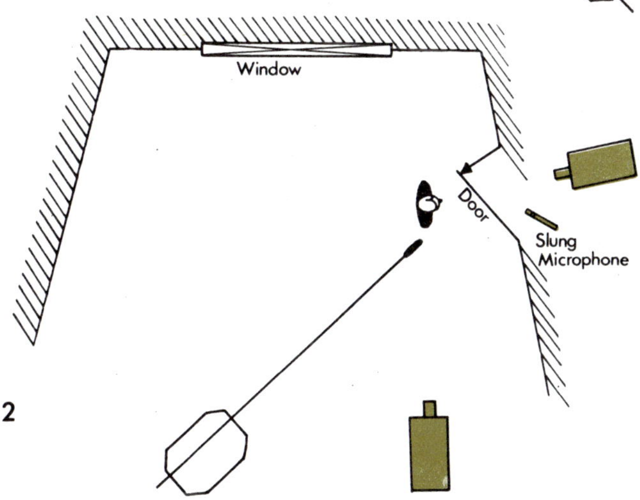

2

PILLARS AND DOORWAYS

Microphone placing

1, If possible the boom should work on the same side of the pillars as the artist. If this is not possible, for lighting or other reasons, two booms may be necessary.
2, When an artist works at a door, or shots are taken through it, it is usually necessary to employ another microphone outside the set. This could be a slung microphone.

Vehicles

Television drama often takes place within the confines of vehicles.

Railway carriages

Railway carriages are usually small cubicles with parallel sides. To overcome the problem of taking shots of the actors on both sides of the carriage it is common to have the cameras looking through gauze pictures or trap doors above the opposite people's heads on either side. This and the frequent requirement to take overall longshots to include the window with moving back projection or colour separation overlay (see page 110) to give the illusion of movement, tend to make it difficult to keep the microphone out of shot. Nevertheless a boom is usually the best answer, possibly with a super-directional microphone (see page 22) to make up for the excessive working distance. The resulting rather 'boxy' acoustic tends to add realism. The sound effects, which are supplied by tape or disc, help to cover any increase in noise level. These have to be carefully synchronised to match the apparent speed of the train, and increased in volume if the compartment door is opened.

Lifts

Lifts present little problem when shot in the usual way from eye level as though from the door. There can however be a temptation to shoot them from above via a mirror to give a claustraphobic effect. In these circumstances it may be necessary to resort to personal microphones on each of the artists. They should be equipped with radio transmitters if they have to move but care must be taken that they do not interact with each other in such proximity.

Motor cars

There is a frequent requirement to shoot action in vehicles of various types, particularly police cars. Usually the cars are seen against back projection (or a CSO screen, see page 110) seen through the side or rear windows, sometimes both. Obviously the sound effects have to synchronise exactly with the apparent movement of the car and the gear changes etc. as well as relate to the other traffic that can be seen. There is a considerable advantage if the film or pre-recording used for the background can be shot with synchronous sound otherwise great skill is required of the gramophone operator. In either case the sound effects should be fed to the actors on a loudspeaker to assist them to synchronise their gear changes etc.

Picking up dialogue from the front seats is fairly simple with a boom if the windscreen has been removed. Otherwise a useful expedient is a small electrostatic microphone mounted on a strip of flat springy material such as laminated plastic which can be sprung into position between the sides to fit flush with the roof.

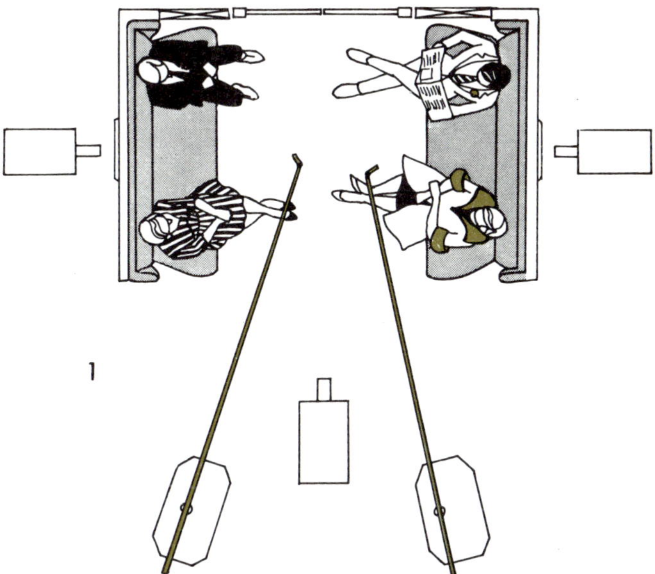

1

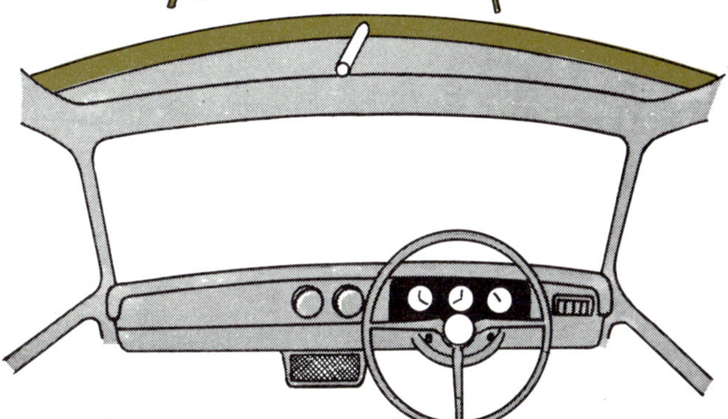

2

IN-VEHICLE SHOOTING

Microphone placing

1, Typical setup for railway carriage effect. The two booms have to work fairly high to clear camera shot through gauze windows above artists' heads. 2, A laminated plastic strip is sprung into the roof lining of a car to suspend a microphone. A loudspeaker may be necessary if 'intercom' is required with a remote 'base station'.

Back projection can produce problems due to noise, whereas colour separation overlay can sometimes make microphone positioning easier.

Back Projection and CSO

Some scenes, notably those which include an exterior background are either too complicated to fabricate in the studio, or are required to give the effect of movement, in which case photographic or electrically generated backgrounds are used.

Back projection

When back projection is required the action takes place in front of a large translucent screen through which a photographic image is projected from a still slide or motion picture film projector.

The sound problems in back projection usually stem from the noise of the film projector. Obviously it helps if the projector is properly 'blimped' but even so whenever possible the whole apparatus should be built into an acoustically treated box.

In designing the set-up for back projection the projector should be arranged to be as far as possible from the acting area and, if possible, heavy velour curtains draped to form a screen between them. Some BP projectors are equipped with an outlet of sound (usually optical). This is probably not of very good quality, but may be good enough to provide sound effects synced to the film. Alternatively the optical sound track can be used to help in syncronising a separate magnetic track, run in telecine.

Colour separation overlay

In modern practice back projection has been largely superseded by an electronic process called colour separation overlay. In this system the area into which the background is to be inserted is brightly lit and painted in a pure primary colour, usually blue or yellow.

An electronic switch is incorporated in the output of the camera providing the foreground and matters are arranged so that wherever this camera sees this colour its output is automatically exchanged for another source such as film (from telecine) or the output of another camera, either live or prerecorded.

In this way the artists, and any foreground objects not of the CSO colour, are inserted into the background (i.e. unlike a simple superimposition, the background cannot be seen 'through' the foreground).

CSO does not normally present problems for sound and, in some circumstances could be used as a positive asset. For instance if sound is required from the 'background' source it may be possible to find a place for the boom which, though well in shot on the background camera is not seen in the final inlayed picture.

Care must be taken to match the acoustic and environmental effects of the foreground and background sound to the composite picture.

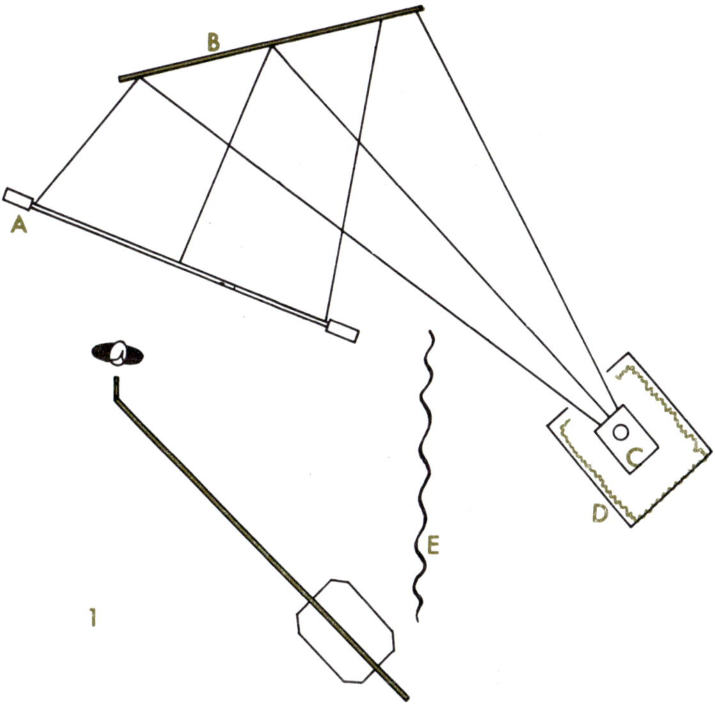

1

2

PROJECTED BACKGROUND

1, Back projection
Typical setup for a back projection sequence. A, BP screen. B, mirror (folding-back the beam to accommodate the length of throw). C, projector. D, sound absorbing box. E, heavy velour curtains.

2, Colour separation overlay
(Left to right): The foreground picture is shot against a background of the CSO switching colour. The background picture is composed to fit the switching colour area. Note stand microphone in shot on the camera. In the combined picture the foreground has overlaid the background and the microphone is now hidden.

Weather and Sound Problems

Wind

Most sequences shot out of doors are subjected to wind. Pseudo-exteriors shot in the studio can be made to look much more realistic if the foliage and the artists' hair and clothing are blowing about. Unfortunately wind on the microphone does not sound like wind at all. Usually it is more like thunder.

The first essential, therefore, where wind is likely to be encountered is to have effective windshields on all microphones. For this purpose the small close-talking shields, usually made of plastic, supplied with many microphones are of little use. To be effective a windshield needs to be at least 10 cm. in diameter with a smooth, round or streamlined shape. It must be mounted firmly to the microphone within its suspension so that the microphone cannot move inside it.

Wind machines

When wind machines are used, care must be taken in planning for the type of machine and its location to localise the wind and keep it off the microphone.

The so called silent wind machines which use a multi-bladed variable-speed fan mounted on a wheeled trolley can, with care, be used for sound sequences. When only a close-up shot is required sufficient wind over a small area can be obtained from a small domestic type fan, mounted on a stand or hand held.

Trick lines

Where large movement of foliage is required it is always worth considering whether a more realistic effect could be obtained by using a 'trick line' to shake the branches. In any case the effect should be augmented, and any machine noise covered, by a recorded effect of wind noise.

Rain

Where 'practical' rain is involved there are two main problems for sound. One is to prevent the microphone getting wet. This can be achieved by enclosing the microphone in a thin plastic bag within a wind shield. This is to prevent the sound of the rain hitting the plastic bag. Surrounding the microphone with a continuous thin membrane in this manner tends to destroy its directional characteristics and can result in turning a cardioid microphone into an omnidirectional one but should not otherwise affect its response.

The other problem associated with rain is noise, particularly the noise of water dripping from roofs and sills. It is advisable to use sacking or tightly rolled chicken wire to break the fall of the water, especially if it is supposed to be on the other side of the window.

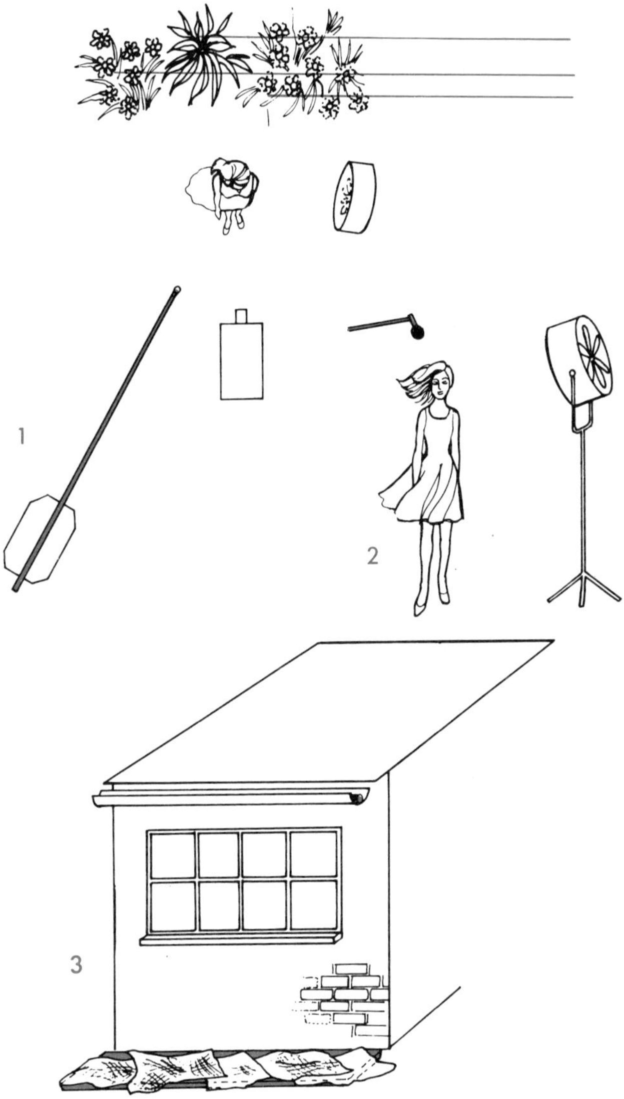

CREATING WEATHER EFFECTS

Wind and rain

1, An 'exterior' scene. Note the 'trick lines' to shake the foliage and the small wind machine to rustle artists' hair and clothing. 2. Front view shows how the fan should be tipped down to prevent wind blowing on to microphones. 3. Roof and window prepared for a rain sequence with tightly rolled chicken wire in guttering and sacking on drip tray on ground.

Gunshots have to be made to sound loud by acoustic quality rather than volume, which is limited by the transmission medium.

Gunshots

Quite a large proportion of televised plays involve the use of firearms which have to be fired with realistic effect.

Real Guns

If the gun has to be seen to fire the best arrangement is obviously to use a real gun with a blank cartridge so that the flash can be seen to synchronise with the bang. This creates a big problem for sound because in many cases the gun is fired in quick succession to the dialogue and is picked up on the same microphone. When the volume has been controlled to keep within the transmission limits, because it has to be, either manually or by a limiter it will come out only slightly louder than the speech. A puny shadow of the original. The best way round this problem is to use a limiter in the output of the dialogue microphone (usually a boom) to 'cut the top off' the peak of volume. The limiter must have a short recovery time (see page 64) so that it does not hold back the succeeding dialogue. It does not matter much if the resultant sound is a bit distorted as this can increase the subjective effect of loudness, as the ear tends to distort the sound when close to a loud bang.

At the same time another microphone placed at some distance, possibly high up in the studio, should be faded up. (With echo added if the scene is supposed to be indoors), to collect the reverberation and give the sound a 'tail' and thereby an impression of loudness.

Don't be afraid to use a full charge cartridge — the louder the sound in the studio the better.

Electronic effects

Where it is not necessary to see the flash of the gun the effect can be laid in with a tape or disc recording or by means of an electronic gunshot generator. This is an electronic device using "triggered white noise" (a signal made up of equal proportions of the entire frequency spectrum in the form of an impulse). This can be set to produce an effective representation of all types of small arms effect and even the characteristic whine of the richochet, by merely selecting the required effect on a rotary switch and pressing the button on cue.

Foldback

Whenever any artificial method of reproducing gunshots is used it is essential that the artist hears the bang as loudly as possible to induce the necessary reaction.

The sound must therefore be 'folded back' on a powerful loudspeaker situated close to the action.

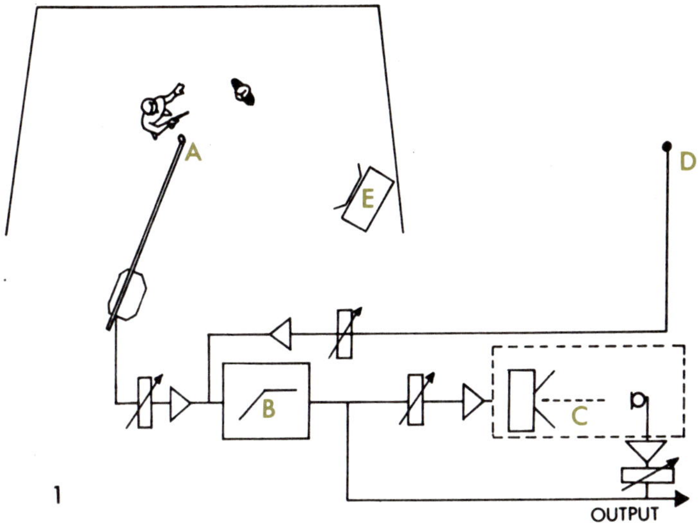

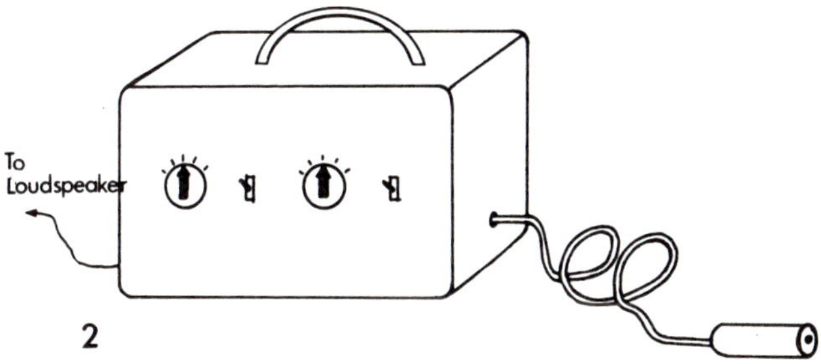

GUNSHOTS

Set-up for murder!
1: A, dialogue microphone. B, limiter. C, echo chamber. D, distant microphone slung high in the studio. E, powerful loudspeaker for use with gunshot generator.
2: General appearance of gunshot generator.

It is an established convention that the distant end of a telephone conversation should sound distorted and at lower volume.

Telephone and Intercom Effects

Telephones are used frequently in drama productions. At the first suggestion that one is to be used the first thing to ascertain is the period and type of telephone required. If it is an unusual variety, e.g. foreign or very old fashioned, it may be necessary to get hold of it well in advance to make it 'practical'. This is because the two ends of the conversation will be in different sets which may be so separated that the artists need the telephone to hear each other.

Telephone distort effects

There is a convention in television drama as in radio that if it is necessary to hear both ends of the conversation the distant end is distorted and lower in volume than the speaker seen in shot. The distortion usually takes the form of a low and high frequency roll-off at about 12dB per octave from below about 500 Hz and above about 1·5 kHz depending on the type of voice. The proper volume for the distant end really depends on the length of the shot. In reality only the user hears the telephone but it does not seem too incongruous if the shot is reasonably close. It seems ridiculous to hear it in long shot.

Effects switching

When, as is often the case, the shots are intercut between the two ends of the conversation it is vital that the sound cutting always follows the vision exactly, even if the vision switches at the wrong time. If you see a person apparently in close-up with 'telephone' sound it is natural to believe that the sound has gone wrong. To ensure that the picture and sound exactly synchronise, a relay-operated effects unit can be made to switch the output of any of four sources either direct or through a variable filter and attenuator to create the desired effect. The switching in and out of the effects can be done manually or tied to the vision switcher by means of a series of keys which select a particular camera to switch a particular source. This can result in split-second synchronising so that it does not matter if the vision switcher cuts in the middle of a word. The appropriate sound will follow provided that the sound supervisor has selected the right camera/microphone combination.

Intercoms

When office-type intercoms are required the most realistic effect can be obtained by making the units 'practical' and picking them up on the dialogue microphones.

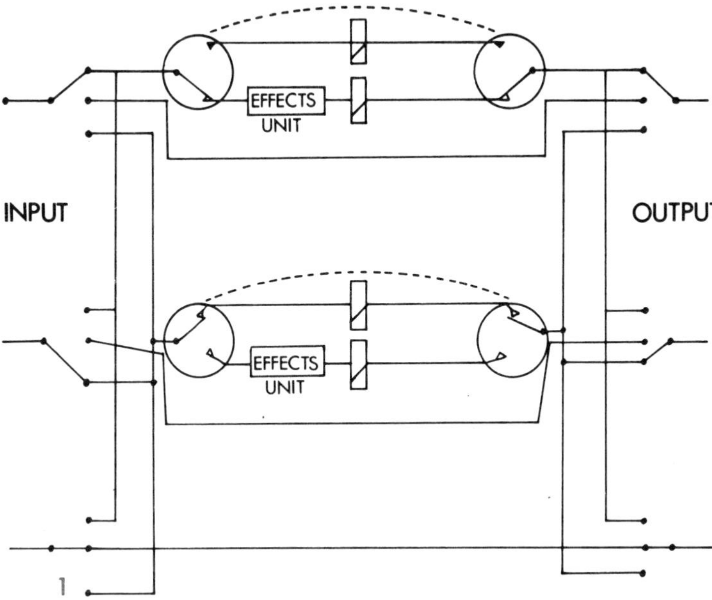

1

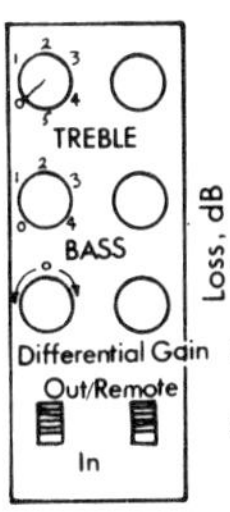

2

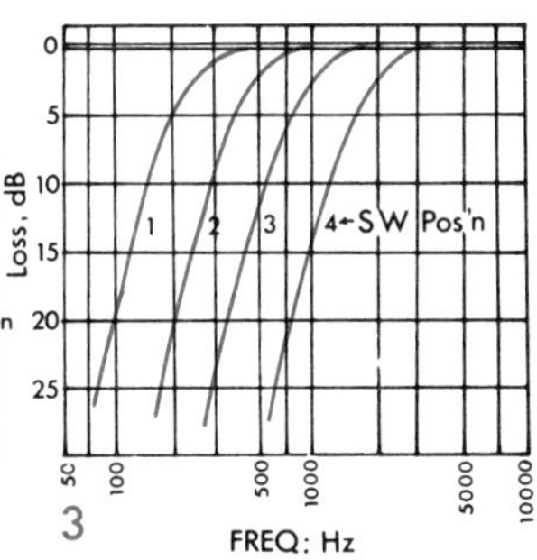

3

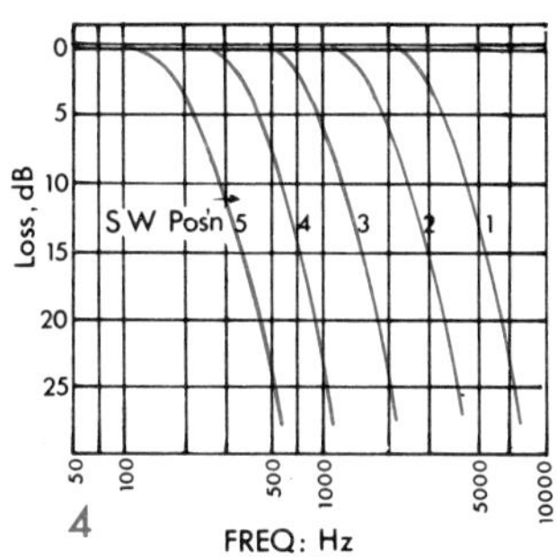

4

TELEPHONE AND INTERCOM EFFECTS

Switching arrangements

1, The switches are relay operated and can be worked either manually or in association with the camera-switching relays. Two effects units are used instead of switching one between the desired circuits so that the levels can be balanced and the optimum effect obtained.

Distant unit

2, A typical telephone effect distant unit, allowing a choice of different degrees of bass attenuation (3) and high frequency attenuation (4).

Exterior Sequences in the Studio

Many dramatic sequences, and sometimes whole dramas, take place apparently out of doors. These exterior sequences often have to be shot in the studio in simulated exterior scenery.

Exterior scenes in the studio

Exterior scenes in the studio are usually shot against a background of a cyclorama lit to resemble sky, or a painted backcloth. In either case this usually consists of a canvas material stretched between poles or runners at the top and on the floor. The main problem is that the outside situation represented calls for an open air acoustic which in real life would be completely dead except for a few discrete reflections from buildings or trees etc. To make matters worse a large exterior usually calls for large, wide-angle shots which means that the booms covering the action have to work further from the artists than normal. As there is usually very little material in the vicinity to absorb, or break up the sound, the effect will sound reverberant. Not at all what is wanted for realism. Matters are considerably improved if thick velour drapes can be hung behind the cyclorama cloth. They absorb sound and also help to stop the cloth billowing in the breeze if a wind machine is used.

The artists should work downstage as much as possible to give an impression of space without requiring too much headroom for the boom.

Use of super-directional microphones

It can be a good idea to use rifle microphones (see page 22) in the booms for these sequences. By including a bass roll-off in their output they can be made more directional, thereby reducing reverberation and making the sound more realistic, as sound tends to lack bass at any distance in the open air.

Footsteps

The other problem is the unrealistic sound of footsteps which can sound just as they are — walking on a studio floor.

If the ground is supposed to be soft it should be covered with a grass mat, real turves, or a thick coating of peat.

If the scene calls for a paved area and particularly if it is on a rostrum where footfalls would sound 'wooden', paving stones should be laid. Real paving stones are heavy and easily broken, so imitation ones have been designed, made of resin and glassfibre on a metal frame which are light and easy to handle and can look quite realistic.

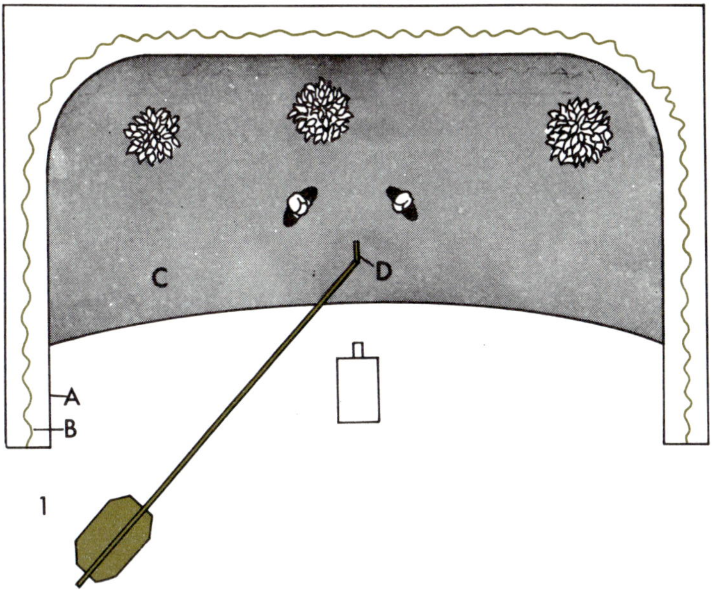

EXTERIORS IN THE STUDIO

Set arrangements
1: A, cyclorama cloth. B, thick velours behind cyc cloth. C, turves or peat on floor. D, 'rifle' microphone on boom. 2: Paving stones (or glass fibre substistute) laid on a rostrum for realistic sound effect.

Short 'rifle' microphones in windshields are useful for picking up sound for film location shooting.

Location Shooting on Film

In spite of the brilliant efforts of scenic designers, few exterior sequences look quite so convincing when shot in the studio as they can when shot in a suitable exterior location. Film is obviously a suitable medium for much outside location work. This may involve producing synchronous sound track, some of which may have to match in with studio sequences.

Sixteen millimetre film is an excellent medium for location shooting because both the camera and the sound recording equipment (usually a ¼-in tape recorder such as the Nagra) are small and light and can be manoeuvred into confined locations.

Synchronising methods

The tape recorder and camera can be synchronised with each other by one of a number of systems. Normally a pulse is recorded on the ¼-in tape either on a separate track from the programme sound or in some cases across the sound track in such a way that it can be recovered separately. These pulses act rather like the sprocket holes on the picture film. In other words they locate a particular element of the sound track with a particular picture regardless of any slipping or stretching that may have occured to the tape.

Later, in the dubbing theatre, the separate magnetic sound recorder is driven from the pulsees recovered from the ¼-in tape so that it is recorded in synchronism with the original picture film.

The pulses can be fed from cameras to tape recorder by means of a wire connection, but if they have to be separated by a considerable distance a miniature transmitter and receiver can take the place of the wire.

An even more convenient synchronising method is to drive the camera and tape recorder pulses from two identical crystal-controlled oscillators so that they remain synchronous even though completely separated.

The microphone techniques for film recording are very similar to those used with electronic picture production. In practice, the most frequently used device is the short rifle microphone (which is usually battery operated) used with pistol grip and wind shield. This makes a simple mobile arrangement that suits many circumstances and matches the 16 mm film camera in manoeuverability.

This is another reason for using rifle microphones for exterior studio sequences (see previous section) if they have to match to film.

Film recordists should always record some 'wild track' i.e. periods of non synchronous 'atmosphere' and any natural noises that relate to the scene. These can be used to give realism in the editing and dubbing process and can be used by the television studio to promote matching with 'live' material.

SYSTEM	SPEED I.P.S	SYNC PULSE Frequency	TRACK POSITIONS AUDIO	PULSE	HEAD DISPLACEMENT & SLIT ORIENTATION	TRACK WIDTH IN INCHES AUDIO	PULSE
PILOTE (MAIHAK)	7½	50 HZ			Erase, Audio rec, Pulse rec, Audio rep	0·2	0·01
NEOPILOTTON (NAGRA)	7½	50 HZ Europe 60 HZ U.S.A			Erase, Audio rec, Pulse rec, Audio rep	0·2	0·03 Total
PERFECTONE	7½	100 HZ			Erase, Pulse rec, Audio rec, Audio rep	0·2	0·01 Each
AMPEX (FAIRCHILD)	15	60 HZ			Erase, Pulse/Audio rec, Audio rep	0·2	0·2
RANGERTONE	7½	60 HZ			Erase, Audio rec, Pulse rec, Audio rep	0·2	0·03

LOCATION SHOOTING ON FILM

Location unit

Top: A film location unit at minimum consists of a camera and microphone. The sound and synchronising track from the camera are recorded simultaneously on a small battery tape recorder. *Bottom:* Sound/camera synchronising system.

'Rifle' microphones mounted on booms, hand held, or mounted on cameras can be used for electronic location production.

Electronic Cameras on Location

The technique of sound pickup for outside broadcast location shooting is very similar to studio work except in three important respects, weather, noise and difficult circumstances.

Weather

The main weather hazard on outside broadcasts is wind on the microphone. This means that in practically every external sequence every microphone has to be fitted with a really effective windshield (see page 38). Similarly it may be necessary to protect the microphones from rain.

Noise

Very few outside broadcasts are totally free from noise. Often the type of noise will be quite incongruous to the production such as aircraft or traffic noises. The noises may be based on a particular frequency, in which case a 'notch' filter or octave band filter can be very useful in recording a period drama. The only real solution is to choose the site and the moment carefully and adopt as close a microphone technique as possible. Then suitable recorded atmosphere and sound effect noise can be added to obscure any incongruous background that cannot be eliminated.

Use of booms

There are some situations in outside broadcasts where it is feasible to use booms and this is certainly ideal for drama where it is practical. For outdoor sequences the booms are usually fitted with 'rifle' microphones mounted in large windshields. The boom can be tracked over fairly level ground but tracks may be needed in rough terrain.

Camera mounted microphones

In some circumstances it can be a useful ploy to mount a rifle microphone on a camera so that the two are aimed together without need of a separate sound operator. This does not work when the camera is using a long-focus lens at a distance from the artist, but is an excellent way of picking up effects noises and crowd scenes etc.

Radio microphones

Undoubtedly the most useful tool for outside broadcast work is the radio microphone. Within the constraints mentioned on page 28 this gives the artist complete mobility. The microphone/transmitter arrangement can also be useful for static microphones to eliminate a difficult cable run, e.g. for effects microphones at the jumps at racecoses.

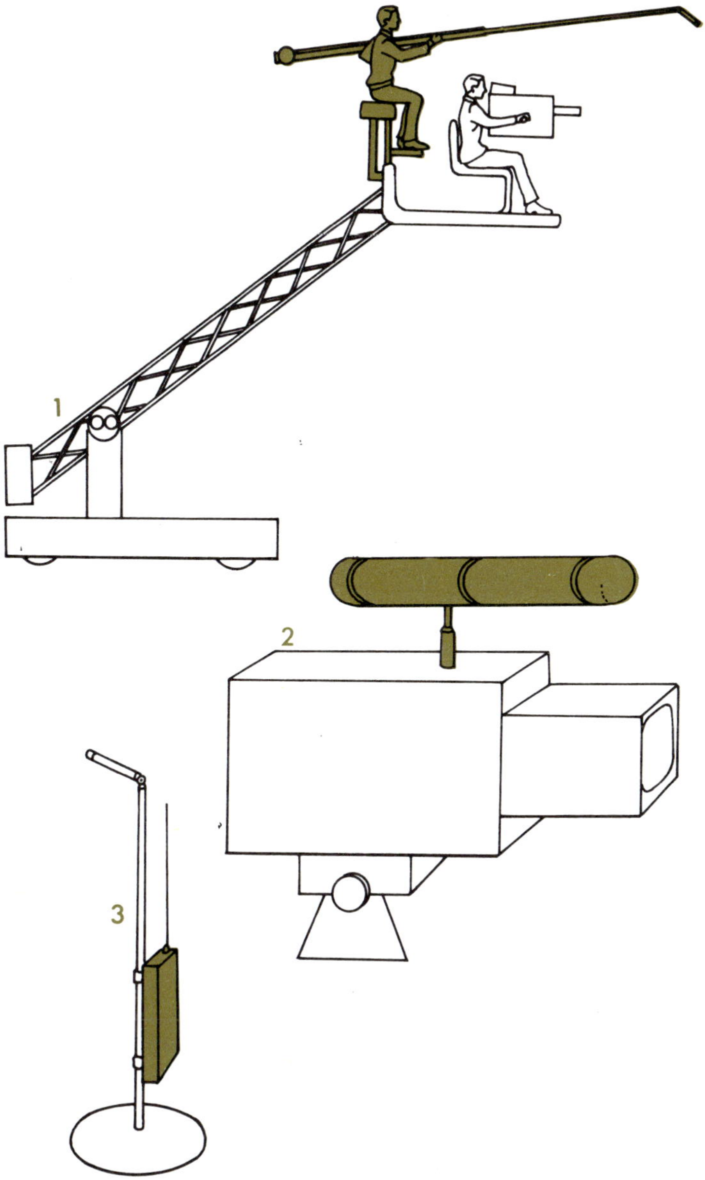

LOCATION SHOOTING WITH ELECTRONIC CAMERAS

Equipment

1, Large camera crane with spigot for mounting boom either with operator or locked off. 2, Rifle microphone with windshield attached to camera. 3, Small microphone and stand equipped with transmitter for relaying sound from awkward locations.

Not all the action noises made by the actors sound realistic; some have to be augmented by 'noises off'.

Sound Effects in the Studio

Most of the noises necessary to give reality to the dramatic scene occur naturally with the movement of the artists, unlike radio drama or most film dubbing theatre practice where the actors stand still and noises such as doors opening, people walking etc. are made by effects operators.

Unreal 'props'
Some of the settings and props are not made out of the material they are supposed to represent and therefore make the wrong noises e.g. safe doors and 'iron girders' are made out of wood, and rocks out of plastic (possibly polystyrene). When the action would result in such objects making a noise it is necessary to use recordings of the real thing which have to be timed to match the action very accurately.

Off stage sounds
Some effects sounds have to occur out of vision, such as a distant door slam etc. This requirement should be foreseen in the planning stages and a special effects door (or other prop) obtained and placed at the right distance from the dialogue microphone to give the required distant effect. It is worth noting that many of the theatrical conventions for 'noises off' such as shaking metal plates for thunder or dropping piles of scene braces for crashes rely upon their loudness for their effectiveness. When relayed at low volume through a television set they sound just what they are. It is much better to use recordings of real noises.

Bells
Quite often the action calls for the ringing of a bell or buzzer on cue. For the purpose the assistant floor manager can be provided with a bell box — a portable device containing a number of bells and a buzzer with push buttons and a battery. The position of the bell box must be set in relation to the dialogue microphone to achieve the right acoustic effect.

Telephones
The best way to make a telephone ring sound natural is to ring the actual phone on the set. This is obviously the right sound and it can be given the right acoustic relationship to the dialogue. The telephone can be rung by means of a special telephone ringer. This gives it the proper mark/space ratio (burr-burr) but arranges that it always starts the first ring instantly on cue. The telephone ringer is also connected to any other telephones used in the action and provides the necessary polarisation so that the actors can hear each other over the telephone when they are working in different sets. The device can also simulate manual or foreign ringing and provides an output of dialling tone.

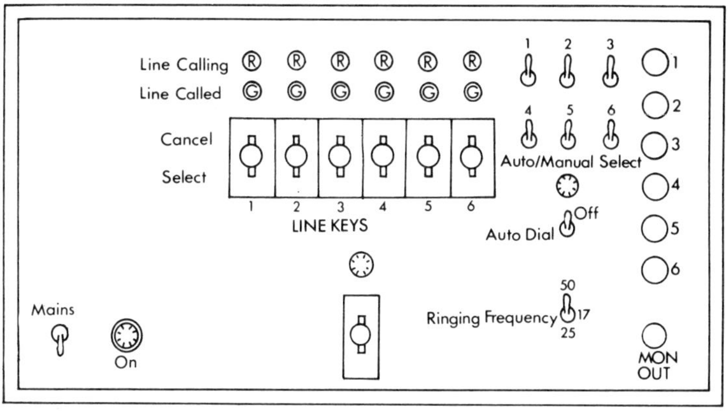

1

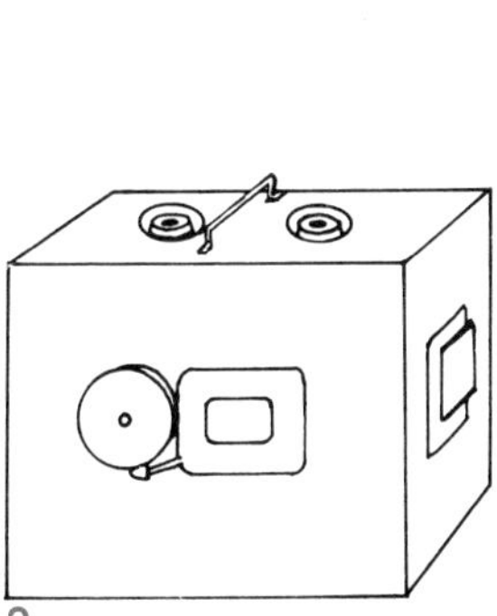

2

SOUND EFFECTS IN THE STUDIO

Equipment

1, Unit for interconnecting telephone in the studio. It acts like a six-line manual/automatic telephone exchange (the switches at the top right are depressed for telephones with a dial). Telephones to be rung are selected on the 'line called' key (green light on). If the auto dial key is selected, dial telephones can ring the selected telephone by the action of dialling. Picking up the telephone gives dialling tone, which can be used for the broadcast. 2, Typical bell and buzzer box containing batteries and worked by push buttons.

Gramophone Operating

Gramophone records are used extensively in television for music and sound effects. Microgroove records, running at $33\frac{1}{3}$ or 45 rpm are used almost exclusively nowadays.

Groove locating

Whatever the type of record it is necessary to start at the right point whether it be the beginning or some particular phrase precisely on cue.

It is therefore usually necessary to start a cue with the record stationary and the stylus already in the groove close to the cue point.

Turntable mechanism

Professional turntables usually provide a type of groove indicating device consisting of a numbered transparent scale attached to the pickup arm. A light is projected through the scale on to a ground-glass screen. As the pickup arm rotates, the appropriate groove number is shown on the screen. There is a lever mechanism to drop the pickup gently on the record. Gram desks are provided with a 'pre-fade' output so that the operator can listen for the cue without fading it up, so that the output is not heard by the audience.

Finding cues

The method of cueing in discs is to play the recording on pre-fade until the cue is heard and then stop the turntable by means of the lever just described. The turntable is then rotated backwards (slowly so as not to cause the stylus to dig in) for about ¼ turn. When the record is cued the turntable is started and the disc should be fully up to speed by the time it is faded up.

Effects discs

BBC effects are nowadays recorded on 7in (18cm) discs at $33\frac{1}{3}$ rpm. They are arranged with the most important cue at the beginning of each band. At one quarter of a revolution before each sart there is a 'pip' of tone. This is a transient pulse which dies away slowly which has been recorded in reverse so that the sound builds up to an abrupt stop to give a precise cue.

Slip mats

If the above method is not sufficiently slick the cue can be speeded up further by interposing a 'slip mat' between turntable and record. The record is then held stationery with a finger so that it slips on the mat with the pickup in the groove. When it is released it will start almost instantaneously if the frictional properties of the mat have been well chosen.

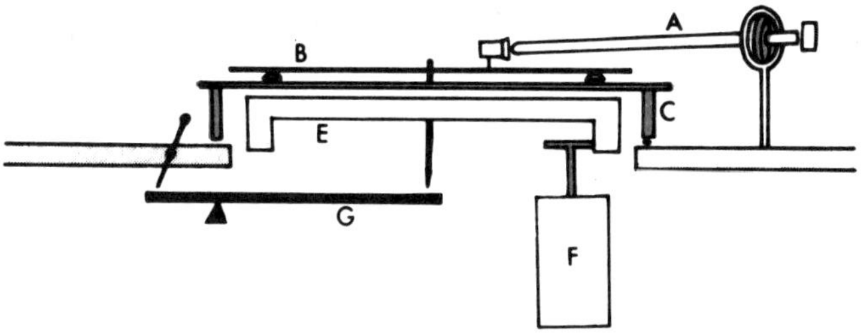

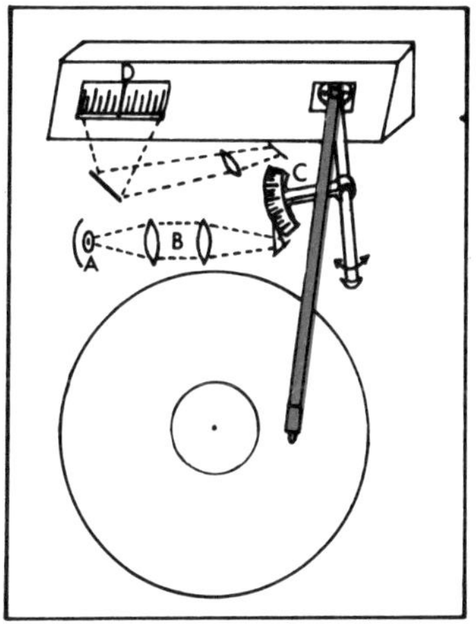

GRAMOPHONE OPERATION

Dropstart mechanism
Pick-up A, with stylus in groove of record, B, supported on rubber bushes on aluminium plate C, which rests on tripod legs on the baseboard D. Underneath the heavy turntable, plate E is spinning, driven by the motor F, when the start lever C is moved, the turntable is raised to meet the aluminium plate which comes up to speed within an eighth of a revolution.

Disc groove indicator
The lamp A is focused by a lens system B on to a Perspex plate with a scale engraved on it, C, which pivots with the pick-up arm. An enlarged image of the scale is thrown on to a ground-glass screen which includes a hair-line cursor D. The image moves in sympathy with the pick-up giving an indication of its position on the record.

Tape recorders are used to play in specially-recorded music and effects and for supplying the sound for mime.

Tape Recording

Quarter-inch tape recorders are used extensively in the production of television sound. The most useful type, suitable for most purposes, is the twin track 7½/15 ips (19/38cm/sec) type. Twin tracks are useful for a number of television operations, such as mime. When artists mime or dance to pre-recorded music it can be helpful to provide two different balances, one for broadcast and the other with a particular emphasis on rhythm to help the artist to synchronise.

Sometimes vocalists are pre-recorded (for practical reasons) and the orchestral accompaniment has to be added live. In this case the second track can contain a rhythm beat and bar count fed to the conductor to ensure that the orchestra remains in synchronism through long breaks in the vocal line.

Another use for twin tracks is when effects recordings have to have different characteristics while maintaining the rhythm. An example is shots of a vehicle seen alternatively from inside and outside. The two different-sounding, but related, effects being recorded one on each track and the sound cut between them in synchronism with the shots.

Incidental music recording

A good deal of incidental music and music for mime is recorded initially on 8 or 16 track recorders. This is to enable each section of the orchestra and the vocal to be recorded simultaneously but separately so that different treatments can be applied to each and the mixing to what is called the 'reduction copy' can take place without having to extend an expensive orchestral session. The reduction copy can be a single track recorded across both tracks or have vocal on one track and accompaniment on the other. This enables small adjustments to be made to the vocal sound to match the picture. When music or effects have to be fitted to visual action it is not always possible to judge the length of material required. It is a good scheme to use two tape recorders for reproduction arranging that the sound sequences are on the long side and spliced on to alternate reels so that the next cue can be set up while the other machine is playing. Often long continuous backgrounds are called for in addition in which case there will be a need for a third machine. Continuous-loop cassette machines can be useful for this purpose, but the loop must be sufficiently long to prevent the repetition becoming obvious.

1

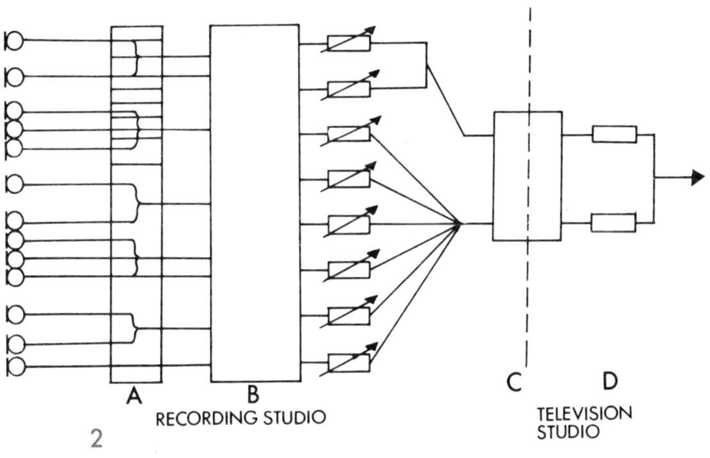

2

TAPE RECORDING

1, Full track and multi-track

The full-track machine uses the whole width of the tape for one track. Although this gives the best signal-to noise ratio, few recordings are now so made in television. Multi-track machines use two or more tracks separated by 'guard bands' ie. small spaces with no modulation. In stereo working it is normal to allow a slightly wider guard band on repro than on record to reduce crosstalk and improve the signal/noise ratio when the recording is played 'mono'.

2, Transmitting multi-track recording

A multi-microphone balance can be mixed A, to record on 8 (or 16) channels, at the recording studio, B, re-mixed (reduced) to 2 channels, C, and finally in the television studio, D, to one channel for transmission.

Tape Editing

Most professional use of tape machines involves editing the tape. This may be to assemble the material into order, adjust its length, or merely to add 'leaders', (blank tape of different colours) to identify start and stop cues.

Cutting the tape

Editing ¼-in tape is achieved by physically cutting the tape. The procedure is to play back the tape until the cue is heard. The machine is then stopped and the two spools rotated slowly backwards and forwards by hand until the exact point of the cue is established. A mark is then made on the tape to coincide with the reproducing head (normally the one on the right) or against a marker post which is fitted to some machines at a measured distance from the reproducing head. The tape is then removed from the head and placed in a special guide, usually made of aluminium, which has a lip slightly narrower than the tape. The guide has a diagonal slit either at 45° or 60° to the tape. The mark on the tape is lined up with the slit, or with a mark on the guide at the same distance from the slit as the marker post is from the repro head, and cut by running a razor blade along the slit.

Joining the tape

The recording tape, or leader tape to which it is to be joined, is laid in the guide and cut in a similar manner. The reason for the diagonal cut is to create a smooth transition and prevent a click occuring due to a sudden change of signal level.

The two tapes are then butted together and about three cms of special adhesive tape is stuck over the join on the back of the tape. Care should be taken to put the adhesive tape on straight and to touch the sticky side as little as possible with the fingers.

Leader tapes

One normal convention is to use white leader tape (about one metre) at the start and to write the title on it. Short lengths (about 15cms) of yellow leader between separate inserts and red leader tape at the end. It is a good idea to get into a routine of marking the tape for splicing in a particular manner, e.g. always along the bottom of the tape. This avoids confusion and the risk of joining the tape on backwards when the splice is made. If it is known that a recording is to contain a number of short inserts but there will not be time to make joints in the tape (e.g. for newsreel) time can be saved by splicing the tape beforehand with rather more tape between the leaders than will be required. The sequences are then recorded immediately after the leader in each case and this forms a marker for each insert.

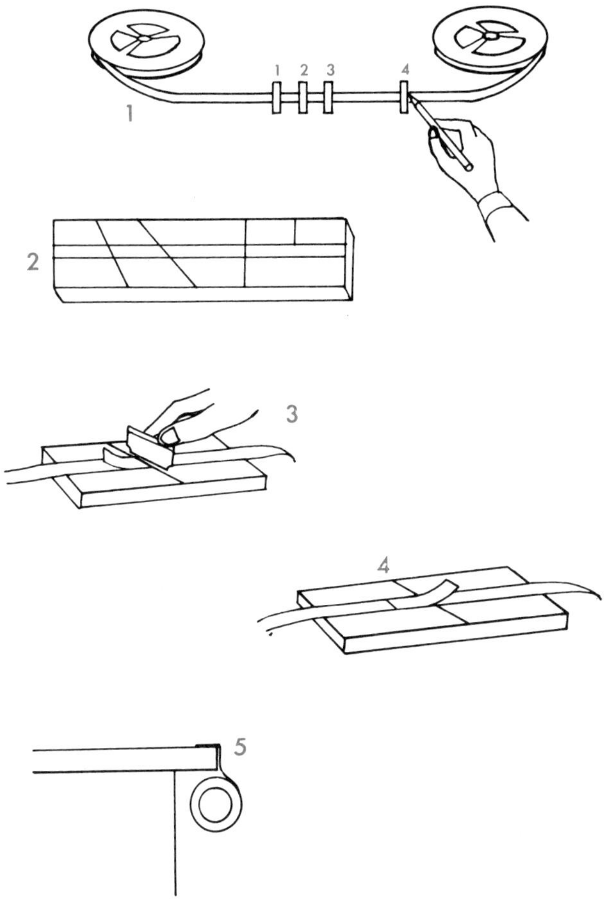

TAPE EDITING

Marking the tape

1, Professional tape-machines have three heads: the erase (wipe) head on the left, record head in the centre and playback head on the right. The editing point is found by 'rocking' the spools backward and forward. It is then marked with a yellow wax pencil either opposite the replay head gap or against a special 'marker post'.

Cutting and joining

2, An editing block with provision for cutting at 60°, 45° and 90°. When the block is used in association with a tape recorder with a marker post a mark is incorporated on the block at the same distance from the slit as the marker post is from the repro head. 3, The tape is cut with a non-magnetic blade together with the piece of tape to be joined. 4, About one inch of joining tape is laid carefully across the join after removing the top cut off piece. 5, Operators usually stick the roll of the tape to the side of the machine for quick action. The area touching the table is cut off before use.

Microphone positions for balancing a piano depend on the quality of the instrument and the type of music being played.

Televising the Piano

Tone production

The piano is a percussion instrument. When a key is depressed a felt covered hammer hits a string and immediately springs away, allowing the string to vibrate. The string vibrates at its fundamental frequency determined by its weight, length and tension, and also at a number of harmonic frequencies which are multiples of the fundamental note. Some of these are discordant with the fundamental and have to be suppressed. This is achieved by giving the hammers a fairly broad surface and covering them with soft felt and arranging for them to hit the string at a point that damps the most discordant harmonics, notably the seventh and the very high 'clang' tones. If the felt has hardened or the keys are struck very hard so that the felt becomes compressed the upper harmonies are excited and the piano has a harsh tone.

Energy distribution

The treble strings are so thin that they make little sound on their own. They are made to pass over a bridge which transmits the vibrations to a sound board from which most of the tone is radiated.

The sound board is less effective for the very high harmonics which promote brilliance and attack. The nearer the microphone is to the strings the more brilliant is the tone. The bass strings, being thicker, radiate quite strongly at right angles to their length. One way of discriminating against a too heavy bass in a 'distant' balance, is to place the microphone toward the tail end of the piano, looking toward the keyboard.

Choice of microphone position

The microphone position for a piano is determined by:

1. The ratio of attack to warmth of tone required to suit the music and character of the performance.
2. The tone of the instrument and its internal balance.
3. The acoustic conditions and possibly the need for separation from other instruments or noise.

The lid should always be open or off otherwise the piano will produce a rather 'stifled' sound. In general the best results are obtained by positioning the microphone so that it can 'see' all the strings at a distance determined by the ratio of attack to warmth of tone required. This can usually be achieved along a line running at about 45° in each plane from the strings.

Generally speaking the harsher the tone of the piano and the more percussive the performance the greater should be the distance.

The optimum position can best be found by 'searching' with the microphone on a boom during rehearsal. When this has been found a neat slung microphone can be substituted.

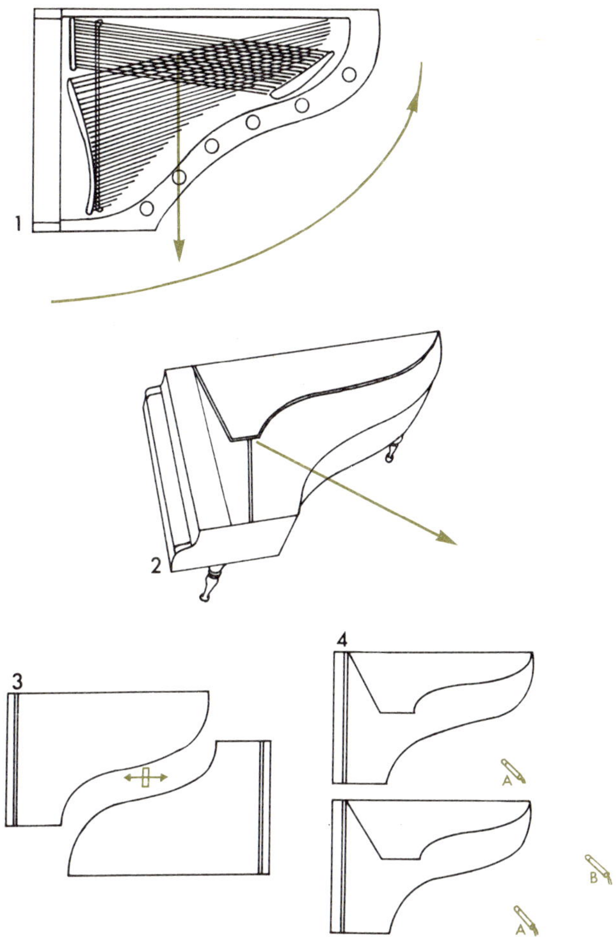

TELEVISING THE PIANO

Microphone placing

1, A concert grand piano can have considerable bass radiation at right-angles to the strings. A more even balance can often be obtained (with a microphone at about four metres distance) by moving it toward the tail. 2, Best results are usually obtained with a microphone on a line diagonal to its three major dimensions so that it can 'see' all the strings. 3, Two pianos 'wedded' for a duet. The lids have to be removed for the arrangement. 4, Two pianos in line formation with balance for two microphones AA, or one B.

Televising the Rhythm Piano

The technique for broadcasting the piano in light music and dance bands is quite different from that used for classical music.

The piano in a dance band

Traditionally the piano has been part of the rhythm section of the dance band and its function has been mainly to provide rhythm chords, filling in phrases, mostly in the treble and occasionally to take the melody. Modern orchestration tends to be rather more complex but the requirement remains for a very crisp brittle tone (particularly at the top end) and good separation from the rest of the band so that, even when played quietly, it can be heard in competition with the much louder brass and saxophone sections.

Microphone techniques

The necessary tone is best achieved by using a unidirectional (cardioid) microphone pointing either almost vertically toward the top strings at a distance of about 20cm or pointing straight down into one of the holes in the iron frame (normally the second from the treble end) and only slightly above the frame so that it looks straight at the soundboard. This latter position does not produce quite such a crisp tone as the microphone pointing at the treble strings, but it can produce a remarkably well balanced result. The reason for this is probably due to the fact that, although the sound board is designed to radiate all frequencies in phase, there is inevitably some antiphase cancellation at the upper end of the scale where the wavelengths are small. Concentrating on a small area tends to eliminate this effect.

A very neat arrangement can be made for either application by using an electrostatic microphone with short extension tube as a table stand resting on a foam rubber pad inside the piano.

Multitone (jangle box) pianos

Sometimes dance band pianists are required to produce a particularly harsh 'jangly' tone (like a coarse form of harpsichord). This is achieved with a multitone piano. Multitone pianos are usually uprights. They have a third pedal (in addition to the sustaining and soft pedals) which causes a set of little cloth strips fitted with metal ends to come down between the hammers and the strings. This greatly increases the harmonic content of the sound and produces a crisp metallic tone.

Microphone positioning

The best mic position for this, and most other upright pianos, is with the top lid or front cover off pointing at the top strings. If this is objectionable in shot the microphone can be placed behind the piano, pointing towards the top end of the sound be...d at about 20cm distance.

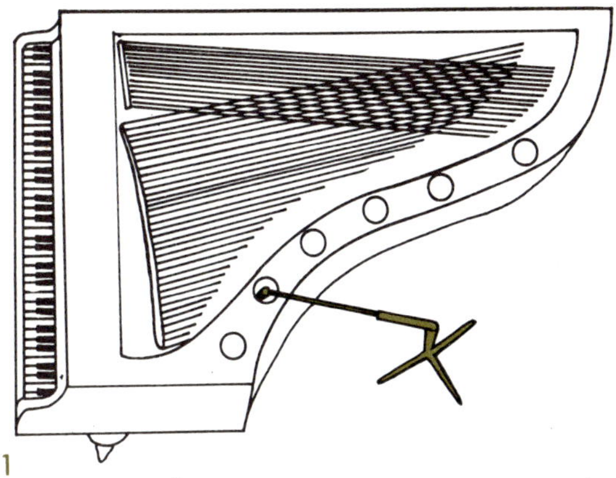

1

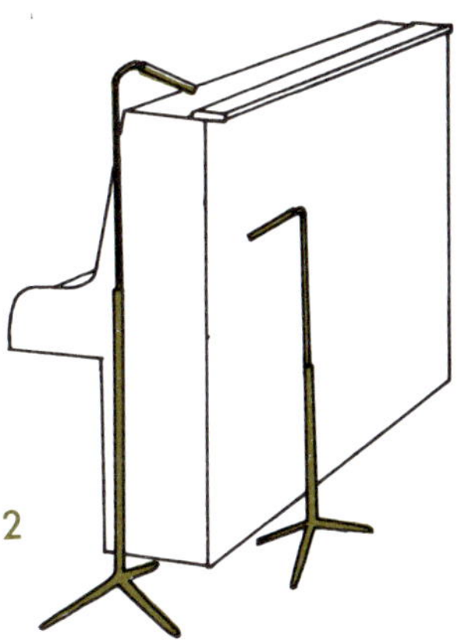

2

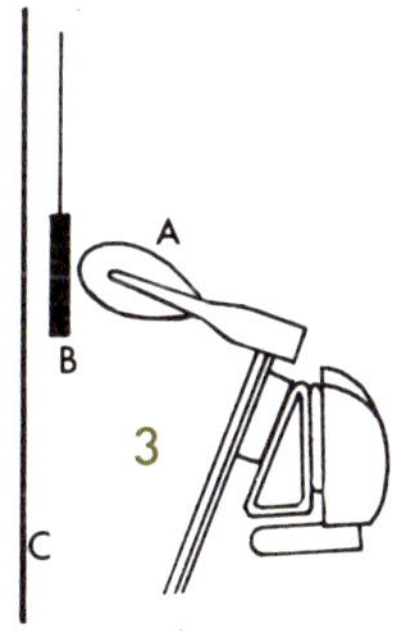

3

RHYTHM PIANO

Electrostatic microphone for close rhythm balance

1, The microphone head would be just above the lip of the hole in the frame. A neater arrangement can be made by resting a short stand on the piano frame or clipping it inside the case. 2, The best arrangement is with the microphone over the open lid at the treble end. Alternatively, it can be placed behind the sound board.

Arrangement for multi tone piano

3: A, felted hammer. B, cloth tab with metal end, C, piano strings.

Various Keyboard Instruments

Apart from the piano there are several keyboard instruments likely to be encountered in television productions.

Harpsichord
The harpsichord is rather like the piano except that the strings are plucked with quills instead of struck with hammers. This produces a small tone that is very rich in harmonics. These harmonics are largely centred on the frequencies where the ear is particularly sensitive so that the harpsichord sounds much louder than a programme meter measuring its output would lead one to believe.

A particular feature of the harpsichord is its loud action noise which stems directly from the plucking action.

The best method of broadcasting the harpsichord is usually to employ two microphones, one underneath pointing towards the soundboard, but angled to discriminate against any noise from the pedals and another microphone above in a similar but rather nearer position than that normally used for a classical grand piano balance.

Clavichord
The clavichord is a smaller, simpler version of the harpsichord. Normally in the shape of a rectangular box, it sits on a table. The strings are struck with a brass tangent which remains in contact with them while the key is depressed so that vibrato (small rhythmic changes of pitch) can be applied by varying the pressure on the keys with the fingers. The tone produced is very small and delicate and calls for a very close balance unless the instrument is completely solo; this is best achieved with a microphone overhead and to the right.

Celeste
The celeste is a keyboard instrument in which the tone is produced by bells. Most of the sound comes out of the back which has an open fret covered with thin fabric.

Celestes vary considerably in purity of tone and power. They usually sound best with the microphone about two metres away from and slightly above the back but the need for separation from other instruments usually dictates a much closer balance, often about 20 cm from the back.

Organ
Most of the organs encountered in television productions are electric. Their sound comes out of loudspeakers either contained within the console or in separate units. It is possible in many cases to take a direct feed from the organ to the sound mixing desk but, unless the organ is properly adapted for this purpose, there may be matching problems and it is simpler to place a microphone in front of the loudspeaker.

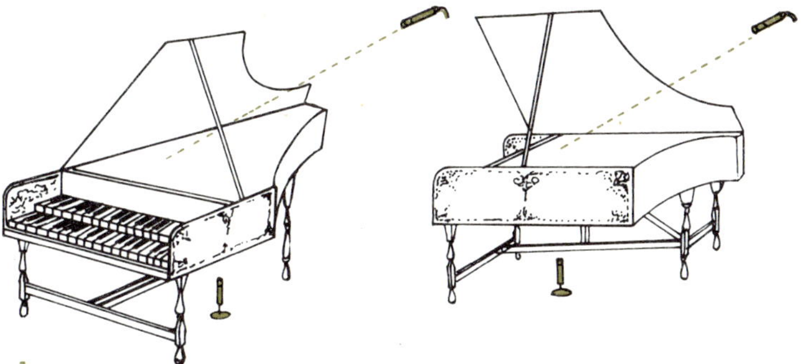

1

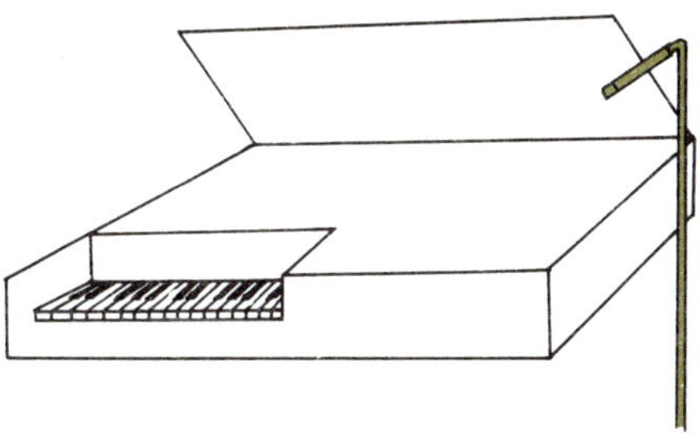

2

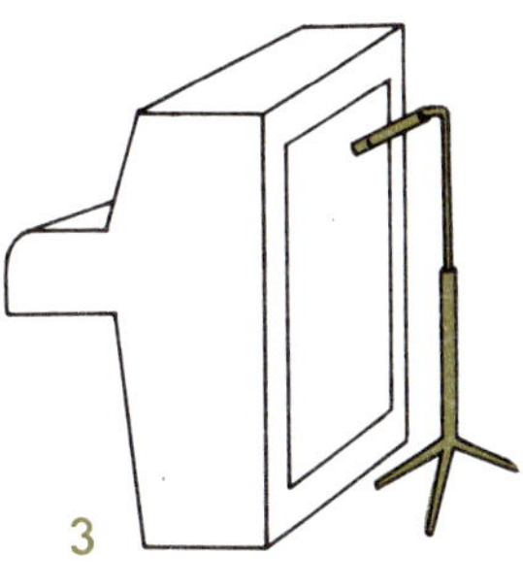

3

4

OTHER KEYBOARD INSTRUMENTS

Microphone positions

1, Two views of a harpsichord showing microphone positions under and over the instrument. 2, Clavicord, showing suitable microphone position. 3, Celeste, with microphone at the back. 4, Electronic organ with alternative position for microphone for internal loudspeaker and external loudspeaker.

Stringed instruments sound best with the microphone roughly perpendicular to their front faces.

Strings

The 'family' of strings used in most orchestras consist of violins, violas, cellos and double basses.

Violin

Most of the tone of the violin is radiated from the front face and from the 'f' holes. The back, which is connected to the front by a peg called the sound post, also makes a powerful contribution to the tone.

The high harmonics which give brilliance to the tone radiate in a narrow lobe at right angles to the face of the instrument.

If a fairly distant microphone technique is used, say two metres or over, the best sound is produced along this line. If, however, a very close technique is necessary to give separation from other instruments it may be better to work somewhat off axis, otherwise the tone can be rather 'whiskery' and variable if the violinist sways about.

The violins in an orchestra sit in pairs (called desks) facing the conductor, sharing a music stand. They are usually divided into two sections of 'firsts' and 'seconds' or, in some types of light music, three groups, A, B and C.

Viola

The viola is a larger version of the violin. It is tuned a musical fifth lower. It has thicker strings and produces a less brilliant tone which is not quite as directional as the violin.

Cello

The cello is tuned an octave lower than the violin. The strings are not unduly thick in relation to its size and it is fairly directional in its upper register. Cellos, or more properly, in the plural, celli, are also arranged in pairs in the orchestra, sharing a music stand between two.

Double Bass

As played in the 'straight' orchestra the double bass does not constitute much of a problem as it is not very directional but it is sometimes necessary to use a fairly close microphone technique to obtain clarity.

Rostrum resonance

Both the cello and the double bass are connected to the floor by the spike on which they rest. This means that the floor acts as an extension of the instrument and if it is a rostrum great care is necessary to ensure that it does not resonate. This is achieved by making a false top with hardboard laid on felt and providing stiff bracing inside. In general the basses sound better on the studio floor.

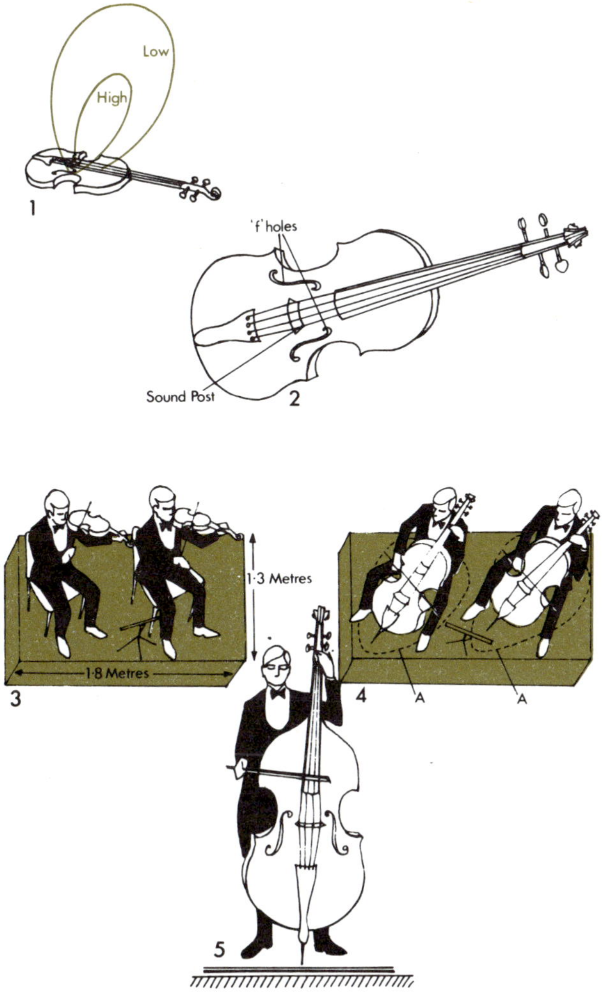

THE STRINGS

Violin tone radiation

1, The violin radiates its high tones and harmonics in a fairly narrow angle slightly forward of normal to the front face. The low frequencies occupy a much broader field. 2, General appearance of violin showing position of sound post and "f" holes. The sweetest tone (from within the resonating chamber) comes out of these holes.

Rostrum space

3, One 'desk' of violins or violas occupies a rostrum space of appproximately 1.8 metres x 1.3 metres. 4, Desk of celli needs at least 2 metres x 1.8 metres unless the musicians sit diagonally. If possible 'cello boards', A, should be used. 5, When basses are used on rostra care must be taken to avoid resonance. The rostrum must be well braced and, if possible, a 'sandwich' of felt and extra top section should be used. A pair of double basses occupies a space of about 2 x 1.5 metres.

Woodwind Instruments

Type of approach

In deciding where to place microphones for the woodwind there are two different approaches according to whether the music is 'straight' i.e. orchestrated for a natural balance in the concert hall or designed for microphone assistance with little if any attempt at internal balance. Most modern and all 'pop' music comes into the latter category.

Flute and Piccolo

The flute and piccolo produce a very pure tone when heard from a distance i.e. the microphone at least two metres away. At close range they have a 'breathy' quality which is often exploited in popular music to produce a crisp effect. In the ultimate the microphone is placed a few centimetres from the mouth piece so that the breath puffs give each note a sharp leading edge.

Oboe

The oboe has a sweet but piercing tone due to the fact that it is rich in harmonics, most of which are centred on frequencies for which the ear is particularly sensitive. The sound that emerges in line with the bell is particularly strident but this is directed toward the floor. Microphone distances of less than one metre are seldom used. The instrument sounds best from a considerable distance. The same is true of the cor anglais (or English horn). This is a larger instrument with a natural key of F, a fourth lower than the oboe. It has a warm complex tone often used to suggest a 'pastoral' atmosphere in programme music. This is created by the unusual shape of the bell which is slightly spherical and closed in somewhat at the end.

Clarinet

The clarinet has a variety of different qualities according to the register in which it is played and the way it is blown. In the high register it can be very piercing. It can also be played 'sub tone' (below its fundamental register) when it sounds very soft and requires a very close microphone technique.

Because of its many registers (usually divided into 'medium', 'chalumeau', 'acute' and 'high') it is difficult to play across the breaks between. For this reason clarinets are made in several pitches, the ones in most common use being B flat and A. Military bands often contain an E flat clarinet which has such a piercing tone it is capable of cutting through the rest of the band.

Bassoon

The bassoons are normally only used in straight orchestras. Their sound is not very directional so they pose few balance problems.

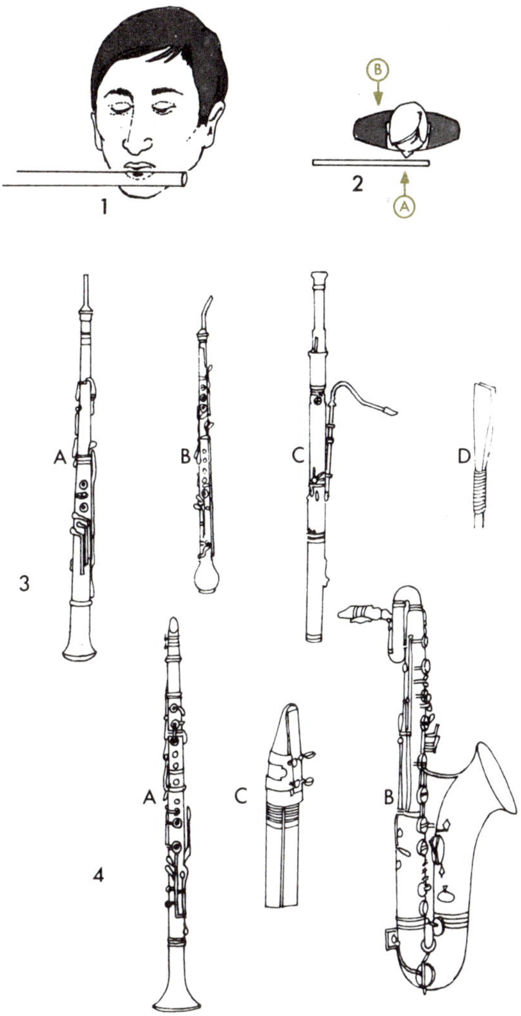

WOODWIND INSTRUMENTS

Tone production

1, The tone of the flute is produced by blowing over a small aperture. (It is said that "you don't play the flute, you stand beside it and pop the tune into it"). 2, A crisp tone is produced with a microphone in front of the mouthpiece (position A; a purer tone from position B). 3, A, oboe, B, cor anglais and C, bassoons use double reeds. D, double reeds are two reeds fixed parallel on either side of the mouth tube, leaving a narrow gap between. 4, A, clarinets and B, saxophones use single reeds. C, single reeds from a narrow gap with the mouthpiece.

Brass Instruments

The brass instruments require a different approach from the woodwind by virtue of their greater power and directional properties.

Trumpets and trombones
Trumpets and trombones project their sound most powerfully in line with their bells. The high harmonic overtones that give brilliance are confined to quite a narrow angle so that if the microphone is right on axis the sound quality varies considerably if the player changes direction. It is, therefore, invariably better to work slightly over the direct line.

When planning the layout for trombone it is necessary to allow space for the extension of the slides. It is also worth considering that the trombones make a spectacular effect when viewed from the side with their slides moving in unison, so they should be arranged in a straight line that is accessible to the cameras.

Mutes
Various mutes are available for the brass. These alter the tone and reduce the volume in various degrees. In general the straight conical mutes produce a hard tone that is not very much less in volume than the open bell (if a microphone is not in direct line). The cup mutes that cover the bell reduce the volume very considerably and produce a softer tone. All mutes, especially the cups, tend to make the instrument much less directional, so that it is satisfactory to work across the microphone.

Horns
The horns require special care in positioning because they point backward almost directly behind the performer so that the surface immediately behind plays an important part in the production of tone. It should provide reflection, but not resonance, and must not be so close that the performer is inhibited from playing at the proper volume.

Euphoniums and tubas
Euphoniums and tubas are mainly used in brass bands and are not very directional; they tend to sound best from an overhead position.

BRASS INSTRUMENTS

Tone production

1, In the brass instruments the tone is produced by pressing the lips against a cup-shaped mouthpiece so that they form a pair of reeds, the air flow being controlled by the tongue. 2, Because of the shape, most brass instruments, particularly the trumpet, produce their most brilliant tone in direct line with their bells (the narrow lobe represents the HF response). Various mutes are used with trumpets and trombones (and to a lesser extent horns). 3, The straight mute which produces a hard tone suggesting distance. 4, The cup mute — a soft smooth tone. 5, The hush mute — a can with absorbent material inside clips over the bell to produce a very soft tone. 6, The Harman, or 'wow wow' mute for shrill 'talking' effect (made by covering the end with the hand). 7, Bowler hat for swell effect.

Small classical orchestras can often be balanced with one microphone.

Chamber Orchestra

The term chamber orchestra suggests a small combination of classical musicians playing music that was written to be played in family intimate surroundings. It may be a quartet or a small orchestra of the type for which much of Mozart's work was written. In any case the chances are that there will be a conventional layout which gives good internal balance.

Pictorial presentation

In keeping with the type of music, chamber groups and small classical ensembles are often depicted on television in realistic period settings so that it would be incongruous to see microphones in shot. Luckily the internal balance is such that in many cases only one or two microphones are required and these can be at quite a distance (say about three metres), where they could easily be hidden behind a chandelier in a long shot.

String quartet

The string quartet has a conventional layout designed to help the musicians to play in concert and balance with each other.

In this type of music the parts are woven together in such a way that sometimes all make an equal contribution to the texture and sometimes each comes into prominence in turn. It should be possible to find a position for the microphone usually about two or three metres in front and above where these conditions are met. If a piano or harpsichord is added another mic will be required.

Classical orchestra

The type of orchestra used by most of the great composers up to about 1800 consisted of about 22 strings in the form 8-6-4-2-2 (1st violins, 2nd violins, violas, celli and basses), double woodwind i.e. two of each (flutes, oboes, clarinets and bassoons), four or five brass (two trumpets and three trombones) two or four horns and timpani.

The standard layout, comprising a semicircle, with the strings in front, woodwind and finally brass and percussion behind makes a logical combination that should balance on one microphone somewhere above and behind the conductor's head.

The procedure is to experiment to find the best position for balance and at the same time the best compromise between clarity (nearness) and warmth of tone (reverberation) then to listen critically to the rehearsal and tell the conductor of any points where the balance is not satisfactory. He will expect this sort of guidance and can usually make the necessary correction.

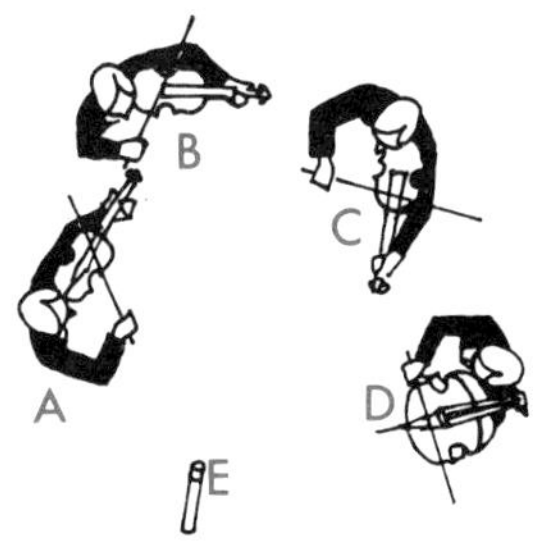

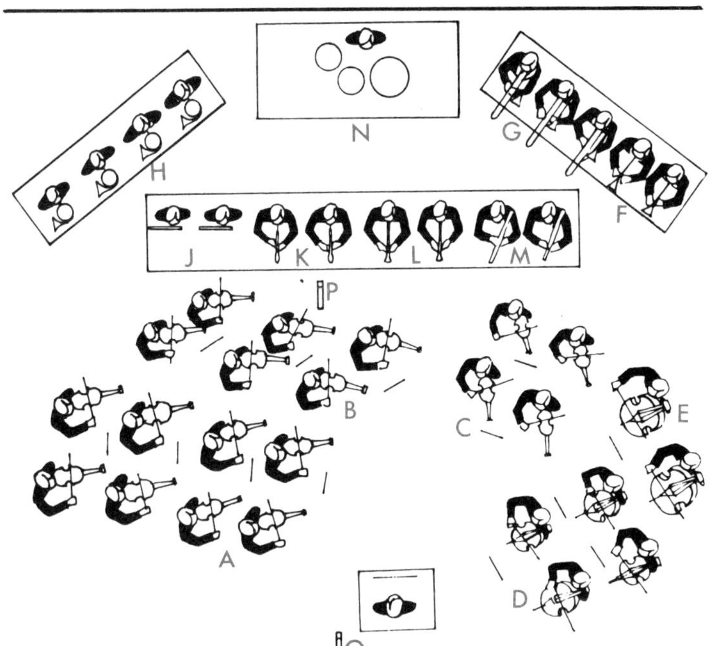

CHAMBER ORCHESTRA

The string quartet
A, first violin. B, second violin. C, viola. D, cello. E, typical position for microphone at about 2-3 metres high.

The classical orchestra
A, first violins. B, second violins. C, violas. D, celli. E, double basses. F, trumpets. G, trombones. H, horns. J. flutes. K, oboes. L, clarinets. M, bassoons. N, tympani and percussion. O, possible position for a single microphone balance. P, position for woodwind accentuation microphone about 2-3 metres high.

The symphony orchestra generally needs several microphones to provide the clarity and attack needed for television.

Symphony Orchestra

The modern symphony orchestra can employ up to about 100 musicians, according to the music to be played. A typical combination would consist of about 66 strings in the form of 20-16-12-10-8 (using the notation explained on page 144), triple woodwind, seven brass, including bass trombone and tuba, four horns and at least two percussion players, two harps and possibly a celeste.

The factor of size

One of the main problems in televising a symphony orchestra is its sheer size. Most symphonic music, with the exception of some modern pieces, is written to produce an internal balance within the orchestra but this only really works in terms of balance, perspective and acoustic effect from a position well back in the hall, or better still, up in the air above and behind the conductor.

The conductor has to get used to a very distorted perspective of the orchestra, standing as he does surrounded by the strings with the brass and woodwind up to 10 metres away. He expects guidance as to the effectiveness of the balance on the microphone. It is possible to balance a symphony orchestra on one microphone in good acoustic conditions and many excellent broadcasts have been made this way. This suggests a distant approach that would be the aural equivalent of a wide angle longshot.

Television perspective

Unfortunately when televising orchestras, directors do not stick to overall long shots because the figures of the musicians would be too small for the viewer to direct his gaze around the orchestra as one does in the concert hall. Instead they track or zoom the cameras in to follow the predominant musical line in close-up. This faces the sound supervisor with a dilemma. If he maintains a distant flat perspective the sound is incongruous with the close-up pictures and produces a frustrating effect, but to follow the visual presentation with matching close up sound would be musically disastrous.

Orchestral balance for television

The solution is to meet the situation half way. To balance the orchestra on a number of microphones, at least three, nearer in each case than for a single mic balance but equidistant in perspective. To these is added the output of a further microphone at a considerable distance; artificial reverberation is used to supply a warm acoustic. This gives overall sound a clear attack which matches well with close up pictures without the necessity to adjust balance for pictorial reasons. It also speeds the process of achieving a good sound and makes it possible to correct to some extent for poor internal balance.

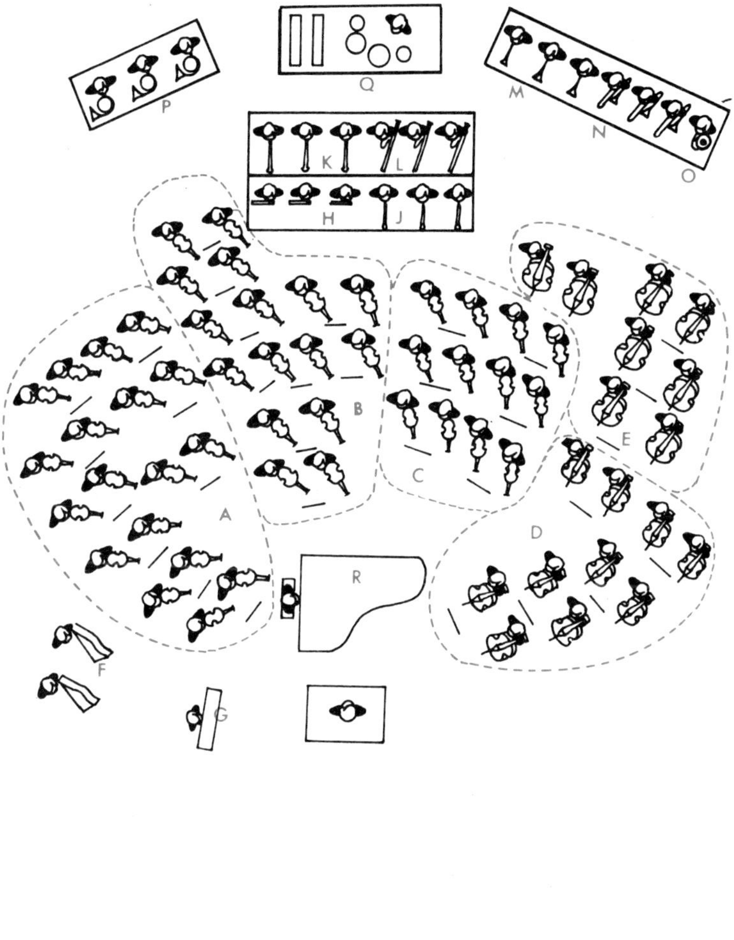

SYMPHONY ORCHESTRA

Typical 'straight' layout for full symphony orchestra
A, first violins. B, second violins. C, violas. D, celli. E, double basses. F, harp. G, celeste. H, flutes. J, oboes. K, clarinets. L, bassoons. M, trumpets. N, trombones. O, tuba. P, horn. Q, tympani and percussion. R, piano if required.

Some conductors favour variations of this arrangement, notably to exchange the second violins' position for the celli. For television, spaced layouts are often employed to give a better opportunity to take pictures of the soloists in the orchestra (especially the woodwind) without giving too much emphasis to the legs of the person sitting behind.

Opera on television can be shot as a theatre relay or, realistically, as a drama with music.

Opera

Televising full scale operas, or even excerpts from operas, can present interesting problems from the sound point of view.

Opera on television is shot largely in close-up with realistic settings. It can play to an audience of many millions most of whom will expect to hear the words. The problem is to obtain clear diction from mobile singers without so subduing the accompaniment that the musical effect is lost.

Studio opera

The larger operas require an orchestra of almost symphonic size and large sets to include many long shots so that the major problems are space separation and time-lag.

The space required to accommodate a large orchestra and even one spectacular set would be enormous. Even if such space were available the power of the orchestra would place restriction on the working distance of the booms and restrict the shots. The orchestral sound reaching the booms, and the singers, would be unbalanced, variable, and come in late, and the singers and orchestra conductor would be unlikely to be able to see each other.

Separate orchestral studio

The best arrangement is to put the orchestra in a separate studio from the action and feed each to the other via loudspeakers. The singers hear the orchestra via column loudspeakers mounted on and tracked with the booms. Then the orchestral sound is always effectively close to them, localised to the back of their microphone and balanced so that orchestral pick-up on the vocal microphone is not so serious. Loudspeakers are also placed strategically in the orchestral studio to relay the singers to the orchestra so that aurally they are close to each other and time lag is eliminated. The singers see the conductor via a camera and monitors scattered about the sets. A repetiteur (deputy conductor) is employed to watch the monitor and relay the beat. With this arrangement the booms can work at a distance of about two metres and a smooth effect is obtained.

Theatre relays of opera

When opera is relayed from the opera house there is less control of the situation. The vocal pick-up is by microphones along the footlights and slung behind the proscenium arch. The rather distant sound is mitigated by the more theatrical presentation and the fact that the action tends to work towards the audience.

A major problem is obtaining a good balance from the orchestra in the pit. This is achieved by placing microphones within the orchestra and adding their output to a microphone slung high over the top.

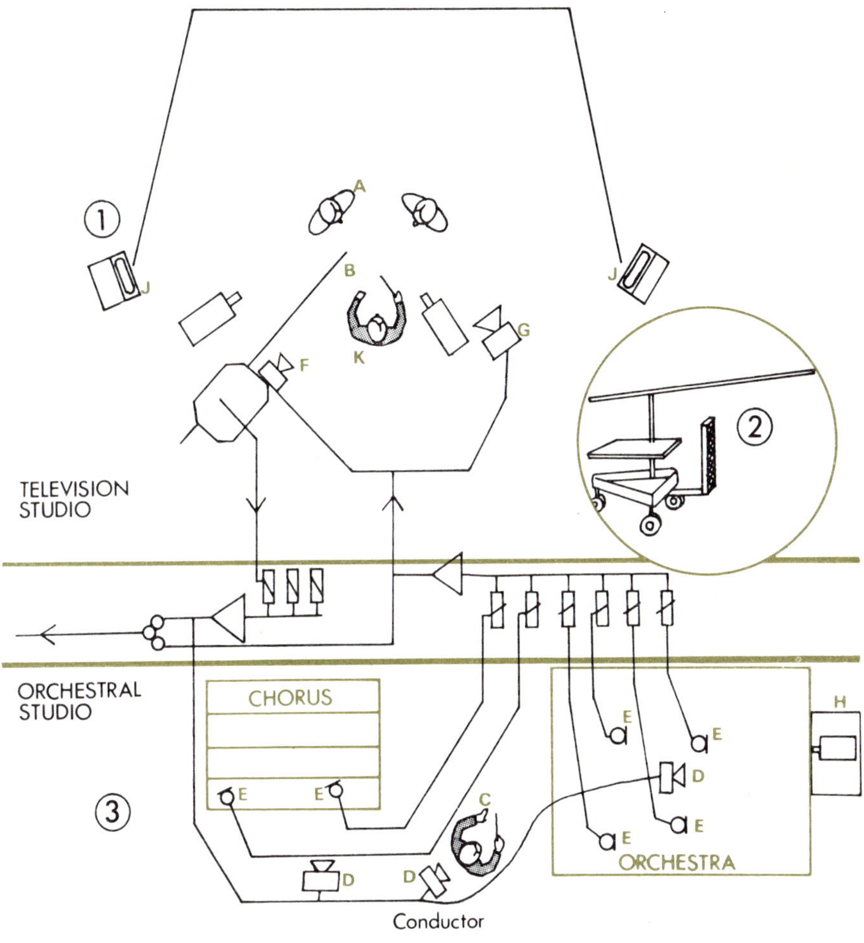

OPERA

Set-up for remote orchestra production of opera
1, Singers, A, in television studio are picked up on boom, B, for broadcast and also relayed to orchestral studio, 3, where they are heard by the conductor, C, orchestra and chorus on loudspeakers, D. The chorus and orchestral sound is picked up on microphones E for broadcast and also fed to artists on special boom-mounted loudspeaker F (see inset 2) and mobile loudspeaker G. The camera H gives a picture of the conductor for slung monitors. J. These are arranged to be on the artists' eye line or the beat is relayed by the repetiteur K.

The Light Orchestra

Light orchestras come in all sizes and combinations and play all types of music from light classical to 'pop'. It is, therefore, not possible to generalise as regards their balance technique except in one respect, they are all likely to require a multi-microphone balance, particularly on television.

Light music balance

A typical combination to play straightforward light music would consist of about 22 strings, typically 8 first violins, 6 second violins, 4 violas, 2 celli and 1 or 2 basses, 3 trumpets, 2 or 3 trombones, 2 horns, double woodwind (i.e. 2 of each; flute, oboe, clarinet and bassoon) some of whom may also double on saxophone, piano, possibly harp, tympani and percussion.

Even if the orchestrations are written for internal balance this is unlikely to be achievable on one microphone. It will be seen that the power of the wind, brass and percussion is very similar to the symphony orchestra but the string section is only half its size. It, therefore, becomes necessary for the microphone to close in on the strings to achieve balance. As a result the strings will sound much nearer than the rest of the orchestra so we have to add equally close microphones to the other sections, fading up just enough of their output to restore perspective while maintaining balance.

Once it becomes necessary to use a close microphone no section must be left without one or it will be missing, or sound very distant.

Multi microphone techniques

The use of separate microphones for each section provides several advantages:

1. Time is saved in finding the optimum position for a single microphone. This can be a premium in the television rehearsal situation.
2. The music arranger is freed from the requirement to consider internal balance and can write for all sorts of novel effects that exploit microphone techniques.
3. Greater freedom of layout. The orchestra can be arranged in a spread formation to help separation or for the sake of appearances.
4. The use of close microphones with added artificial reverberation produces a 'crisp', brilliant tone which suits modern orchestration and is congruous with close-up shots of the musicians.

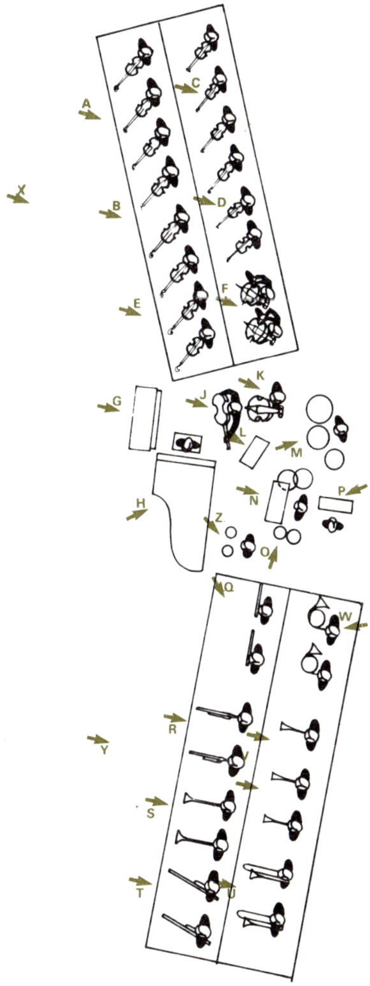

Spread-out split layout for visual effect

Because of the unnatural spread of the instruments, and possibly also to balance for modern orchestrations, a large number of microphones, each at fairly close range, is required. The microphones (pictured here as arrows) accommodate the following sections: Microphones A, B, C, & D, violins. E, violas. F, celli. G, celeste. H, piano. J, acoustic guitar. K, double bass. L, electric guitar amplifier. M, tympani. N, bass drum. O, side percussion. P, vibraphone. Q, flute. R, oboes. S, clarinets. T, bassoons. U, trombones. V, trumpets. W, horns (they project their sound backward). X, overall string microphone. Y, overall woodwind and brass microphone. Z, bongoes. X and Y can be very useful to restore cohesion which may be lost with very close microphone technique. They also tend to bridge the gap between the direct sound and the added reverberation when an acoustically dead television studio is used.

Dance Band

The dance band consists of a brass section (the big bands have 4 trumpets and 4 trombones) a saxophone section (normally 2 altos, 2 tenors and a baritone) and a rhythm section consisting of piano, guitar, bass (string bass or electric bass guitar) and drums.

Rhythm section
The rhythm section is a very important feature of the dance band. In addition to supplying an important element of the sound, which should be reproduced with almost equal volume to the tune, it drives the rest of the band along and gives 'lift' to their performance.

Rhythm piano
The piano is classified as a rhythm instrument in the dance band as a hangover from the early days when the pianist kept up a rhythm vamp almost throughout. Nowadays the piano work is much more subtle, it requires a very close microphone (see page 134) to cut through the power of the band.

Double bass
The bass, whether it be a string bass or electric guitar, should sound almost as loud as the rest of the band. This is simple to achieve with an amplified guitar but much more difficult with a double bass which, when played pizzicato, i.e. plucked, produces a very small sound. It requires a unidirectional microphone within a few centimetres of the strings close to the bridge. Alternatively an omnidirectional microphone can be used softly suspended within the bridge.

Drums
The drums of a dance band have an enormous volume range. Nevertheless, to be reproduced with proper clarity they require a close microphone technique with up to three microphones on one kit. One on the side (snare) drum and hi hat (pedal cymbal), one for the bass drum and possibly one for the tom tom and bongos. This close arrangement ensures a crisp effect to which echo can be added to give the effect of loudness.

If the drums are on a rostrum it needs to be very firm and acoustically damped to prevent spurious resonances travelling up the microphone stands and blurring the sound.

Guitar
The guitar is almost invariably electrically amplified and a microphone can be placed in front of the loudspeakers.

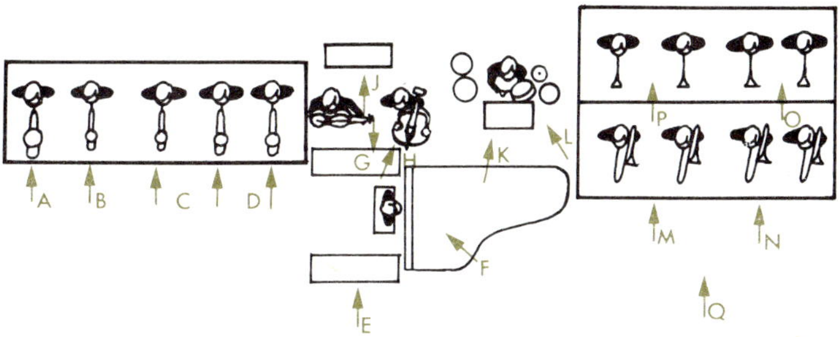

DANCE BAND

Typical television layout
Note the rhythm section in the middle to keep the musicians (who are really too separated for ensemble) in time. The microphones (arrowed) cover A, baritone saxophone (sometimes doubling flute). B & C, alto saxes. D, tenor saxes. E, celeste (if used) or organ. F, piano, G, jangle box piano. H, double bass. J, guitar amplifier. K, bass drum. L, side drums and hi hat. M & N, trombones. O & P, trumpets. Q, overall brass microphone (to give cohesion and a sense of power to the brass sound).

Overhead view of drum kit
Although the bass drum is very powerful it requires a close microphone to produce a crisp attack. Echo can be added to give the effect of power. The side drum and hi hat often involve delicate work with brushes and need a close microphone.

Main problem with accompanying orchestras is achieving sufficient separation from the vocal microphone.

Variety Orchestra

The type of orchestra encountered most often in television studio production is the variety orchestra. This is the type of orchestra that provides the accompaniment for most light entertainment variety and 'music' productions. It can be called upon to play a wide variety of types of music, even in the same production.

The variety orchestra is essentially a dance band with strings and in some cases also a flute, oboe and harp. Like the dance band, the modern variety orchestra requires individual microphones and little attempt is made to produce a balanced sound without them. It is assumed that full use will be made of microphone technique so that it is possible to write a solo for a trumpet with a cup mute (which makes a very small sound) or a clarinet playing subtone (which is even quieter) against the full weight of the rest of the band (see page 142). This assumes not only that there will be a microphone handy but also that the sound supervisor will bring up its level at the right time and down again the moment the trumpeter takes out the mute and joins in with the rest. For this purpose it is a great asset if the sound supervisor has a view of the brass section as each type of mute makes a different volume of sound. Because the string section is likely to be small relative to the brass and saxes it will be necessary to use a split layout and if possible screen the sections from each other to give separation.

Out of shot orchestra

If the orchestra does not come into shot there is much to be said for putting it in a separate studio and feeding its sound to the vocalist on loudspeakers (see page 148). If the orchestra cannot be completely separated it may be necessary to feed the rhythm to the vocalist on a close loudspeaker to prevent the possibility of time lag.

The possibility of pre-recording and miming to foldback should always be considered as this can provide the ideal conditions for achieving a good sound balance.

In shot orchestras

If the orchestra is in shot, a split layout will still be necessary but this can be spread out to achieve a neat appearance with the strings to one side, rhythm in the middle and brass and saxes to the other.

Vocal separation

To achieve proper separation close to the orchestra the vocalist has to work very close to a stand, or hand held, microphone. If excessive movement is required and a personal microphone is an encumbrance it may be necessary to resort to mime.

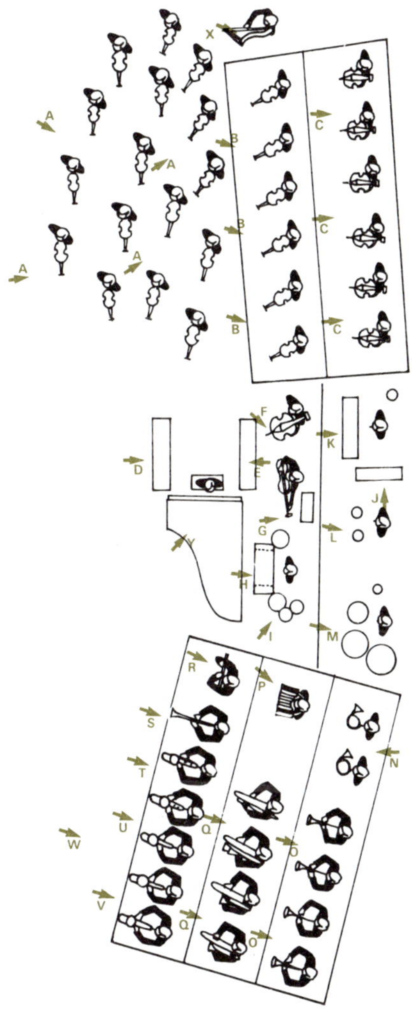

Microphone positions

The microphones pictured as arrows, cover the following sections: A, violins. B, violas. C, cellos. D, celeste. E, multitone piano. F, bass. G, guitar. H. bass drum. I, hi hat. J, vibraphone. K, percussion. L, bonogoes. M, tymps. N, horns. O, trumpets. P, accordion. Q, trombones. R, flute. S, oboe, T, baritone sax, U, alto saxes. V, tenor sax. W, overall microphone for band section. X, harp. Y, piano.

Featured showbands tend to play stylised orchestrations requiring specialised multi-microphone technique.

Concert Orchestra or Showband

Some television productions employ very large orchestras of almost symphonic proportions playing stylised orchestrations that require multi-microphone technique to achieve balance.

These may be used as accompanying orchestras for prestige light entertainment productions involving star vocalists or instrumentalists or as concert performers in their own right. In either case the balance technique is dictated not so much by the lack of internal balance in the instrumentation (in fact some featured orchestras employ as many string players as a symphony orchestra) but by the type of orchestrations used, the characteristic quality of the required sound and possibly the versatility to cope with different types of accompaniment. In addition to a large string section the orchestra will probably have a complete dance band section with five saxophones and eight brass, horn, double woodwind, piano, harp, celeste, possibly accordion and electric organ and certainly a large rhythm and percussion section.

String treatment

Despite the volume of string tone available, many microphones are used, one for about every four desks of instruments (more if the layout is widely spread). This is to ensure a really solid string tone with good clarity and separation from the other sections so that the desired amount of artificial reverberation and possibly frequency response shaping can be applied individually. All the other instruments have practically one microphone for each.

Dance band section

The treatment for the brass, saxophone and rhythm sections is as described on page 152, it being important that the rhythm drums occupy a central location close to the brass.

Visual considerations

If the band is featured in vision, it is necessary to provide good access for camera tracks around the sections including side shots along the rows. All solo instruments such as woodwind, accordion and piano must have good visual as well as aural separation to enable clean close-ups to be taken.

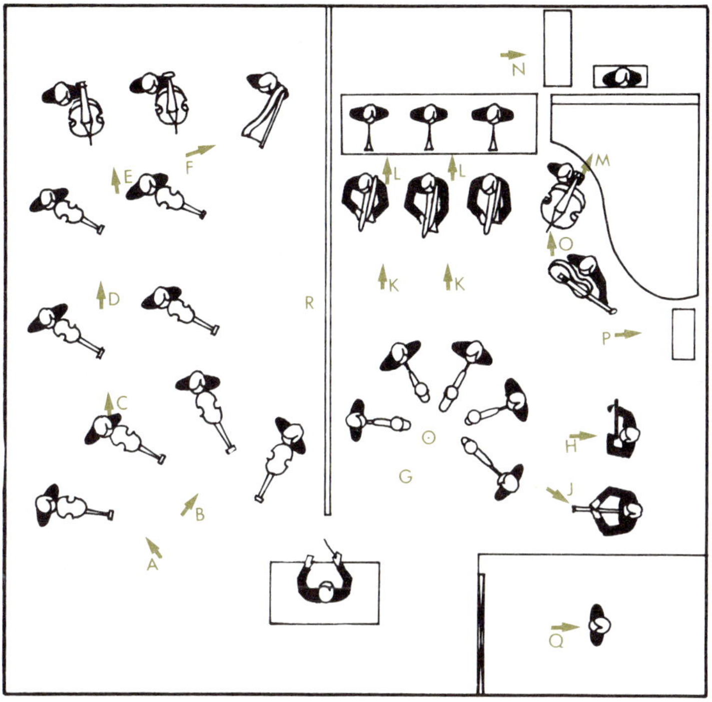

VARIETY ORCHESTRA

Layout and microphones
A compact layout for an out-of-shot variety orchestra in a small separate studio or acoustic enclosure. Microphones cover the sections as follows: A, B, & C, strings. D, violas. E, cello. F, harp. G, saxes. H, flute, J, oboe. K, trombones. L, trumpets. M, piano. N, jangle box. O, double bass. P, guitar amplifier. Q, vocalist in acoustic booth. R, an acoustic screen between the sections helps to balance the weak string section against the brass without preventing all the musicians seeing the conductor.

Pop groups make extensive use of stand microphones and electronic instruments.

Pop Groups

The current 'pop' music scene can be broadly divided into 'Folk' groups, 'Electronic' groups and vocal/instrumental.

The 'Folk' group
In the 'folk' group the tone production tends to be mainly acoustic. They usually consist of acoustic guitars, tambourines, possibly a flute or other wind instruments (e.g. saxophones). All have to be played very close to the microphone to achieve the necessary intimate effect, and as it is customary for some of the instrumentalists to sing and play simultaneously there is often a requirement for more than one microphone for each musician.

'Electronic' groups
At what is sometimes called the 'sharp' end of 'pop' considerable use is made of electronics and nearly all the sound comes out of loudspeakers. Electric guitars, electric organs and sometimes tone synthesizers are used and all non-electric instruments and vocalists are amplified through microphones.

Recording technique
In many cases it makes sense to use a 'backing track' for the accompaniment so as to achieve the necessary separation for the vocal and in order to produce the right effect. This may be pre-recorded on the day, or it may be the original backing used in the production of the artist's commercial record.

There can be an advantage in this because the fans expect the artists to sound exactly like their records which may have taken many hours to make and be virtually impossible to reproduce exactly.

It is normal to record on 8 or 16 track recorders (see page 128). Multi-tracking can be used in which the performers add further vocal or instrumental tracks while listening to their previous recording on headphones. Various treatments can be applied to each of the tracks and many experiments in mixing can be made before they are combined into a reduction copy.

Technical treatment
Most of the tracks have artificial reverberation of various decay times added and the frequency response of each source is individually adjusted. Use is made of 'flutter echo' (discrete echoes, preferably related in repetition rate to the rhythm of the music), usually provided by feeding a proportion of the output of a tape recorder back to its input. Various forms of distortion can be introduced, including the production of square waves to give a coarsening, overloud effect. 'Phasing' is sometimes provided using two tape recorders.

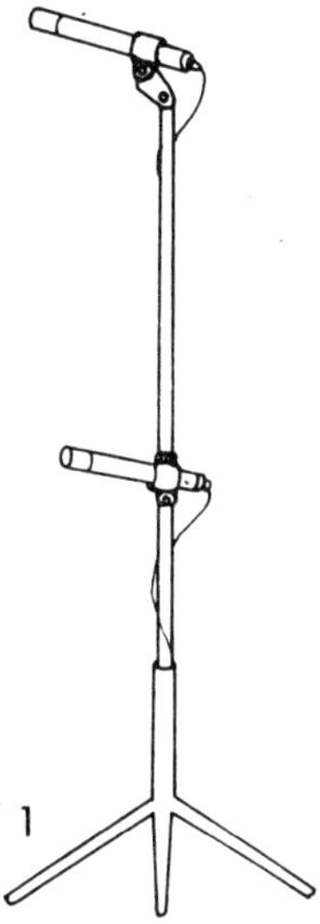

1

2

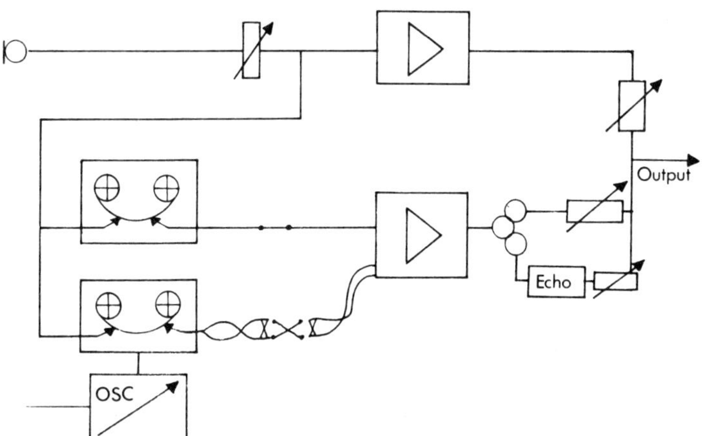

3

POP GROUPS

Obtaining special effects

At the 'sharp end' of the pop scene the sound coverage usually consists of vocal microphones and microphones in front of each loudspeaker. For the more 'folksy' styles two microphones may be needed for some of the musicians — one for the vocal and one for the acoustic pick up of their instruments. 1, a double microphone stand adapter for a vocalist with guitar. 2, typical arrangement for vocalist at piano: electrostatic microphone with short extension on table stand. 3, one method of achieving a 'phasing' effect. Two tape-recorders connected in antiphase, one driven by a variable oscillator. If they run at slightly different speeds the two beat together alternately in and out of phase, creating a sort of frequency-selective tremolo effect.

Sound and VTR

Increasing use is being made of VTR for discontinuous recording and sophisticated editing techniques. It is therefore advisable for sound supervisors to have some knowledge of the VT recording and editing processes to enable them to provide the necessary co-operation and the correct material.

Types of VTR machine

There are basically two types of VTR machine in general use, quadraplex and helical scan. There is also a video disc system which is used (silent) for slow-speed action replay. Most recording for broadcast purposes is done on two-inch quadraplex machines. In this type the vision recording is made by means of a drum which carries four recording heads set at 90° to each other rotating at right angles to the length of the tape. The frequency-modulated vision track is laid across the tape as a series of almost vertical lines. The magnetic particles in the tape are orientated transversely to favour the vision to the detriment of the sound which is recorded longitudinally along the edge of the tape, travelling at approx. 15 ips (38cm/sec) as in an audio tape recorder. There is a second audio track known as the cue track but the quality is not really good enough for broadcasting. It is, however, sometimes used, e.g. for sound effects when they are needed separately from the commentary in sports relays and for address-codes. The address-code provides a unique identification for every picture frame along the tape and can be used for positional selection, cueing and synchronising with other tape machines. There is also a control track used to synchronise the headwheel and capstan motors.

Helical scan VTR

The helical scan machine also uses a rotating drum to increase the head-to-tape speed for vision but in this case it rotates obliquely against the tape which is wrapped around it in the form of a helix.

Unlike quadruplex machines there is no universal format for helical-scan recorders. Various widths of tape are used in various configurations of wrap and head drum arrangements. The FM vision signal is recorded as a series of long diagonal tracks, typically at about 4° to the longitudinal axis, each representing one picture field.

The audio and control tracks are usually recorded longitudinally on opposite edges of the tape as in quadruplex recording. One machine employs a duplex rotating head recording obliquely on 2in. tape. This system also provides two high quality sound tracks, cue track and address-code track.

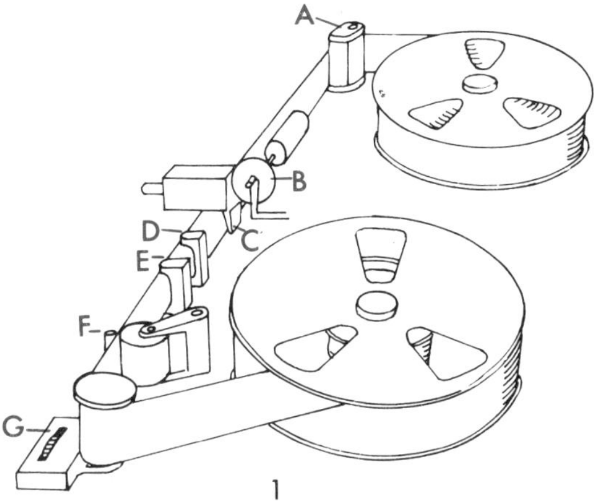

1

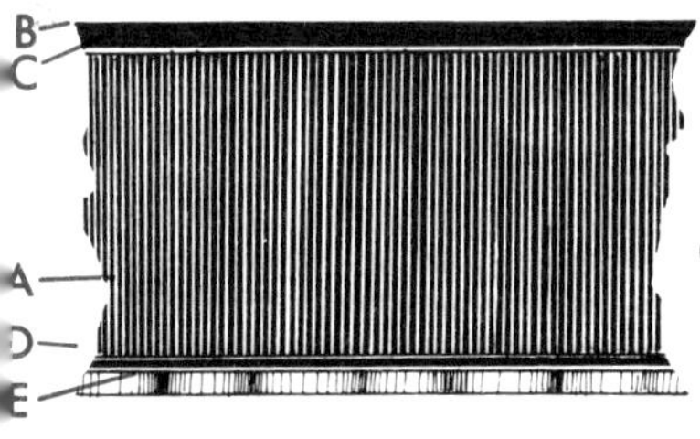

2

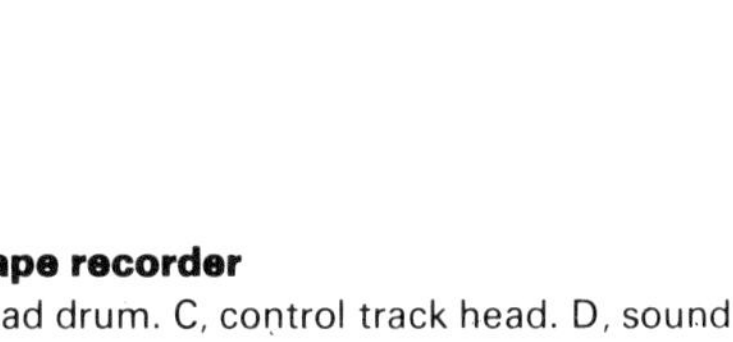

3

SOUND AND VIDEOTAPE

1, General arrangement of videotape recorder
A, master erase head. B, recording head drum. C, control track head. D, sound and cue erase heads. E, sound and cue recording heads. F, capstan. G, timer.

2, Quadruplex recording on 2in tape
A, vision tracks, B, audio track (70 mil) C, 20 mil guard band. D, cue track (20 mil). E, control track (50 mil). This acts as a sort of 'electronic sprocket holes' to synchronise the movement of the tape and the headwheel.

3, Helical scan recording
A, vision track (1 field per track), B, No. 1 audio track. C, control track. D, No. 2 audio. E, cue and address tracks. One helical-scan format incorporates five AF tracks — two programme and one cue track on one side of the vision tracks with the control and address-code tracks on the other.

Sound and VT Editing

Most pre-recorded material involves editing the tape. This might be for a variety of purposes from the simple excision of a mistake to complex 'animation' techniques.

Editing with quadruplex machines can be achieved by physically cutting and joining the tape (by sticking adhesive tape on the back) or by electronic means. In either case, if the recording is made discontinuously, the sound supervisor has to match the levels and quality of the sections, particularly if they are supposed to be in the same sequence, so that they will butt together smoothly. If the matching sections are to be recorded out of sequence or separated by a time interval, careful notes of microphone positions and fader settings may be necessary.

Cut editing

A major problem about cutting the tape is that the sound recording head is approximately 23 cm in advance of the vision recording. This means that if an 'out' cut is made on a vision cue approx 0·6 sec. of the sound of the cut material will remain on the tape. Similarly if the cut is made at the beginning of a vision cue 0·6 sec. of the new sound is lost. Ideally, therefore, there should be no sound for at least 0·6 sec. on either side of the join. If there is atmosphere or effects across the edit it should be non-rhythmic and perfectly matched in level and character. This is extraordinarily difficult to achieve, even for such apparently continuous sounds as audience reaction and applause. Where the sound is continuous the VT editor will 'lift off' the sound by dubbing to an audio tape recorder and re-recording on the the VT over the join.

Use of audio tape recorders

For a simple edit the sound at the start of the sequence to be edited in is copied on to an audio tape and then dubbed on to the end of the previous sequence to coincide with the vision cue. When the tapes are cut and joined together sound and vision should match. If a butt join in sound results in a sudden change in background or level, or if a music link is to be recorded over the edit it will be necessary to dub sound from both sides of the edit point and edit the sound on the audio tape first. A second track (or tracks) can be used to record the music or background effects, separately controlled, to promote a smooth transition of background on either side of the join. The vision edit is then made to match the sound, which is subsequently re-recorded back across the join between convenient in and out points. Sound supervisors can help by providing suitable recordings of 'clean' atmosphere or 'wild track' effects for this purpose.

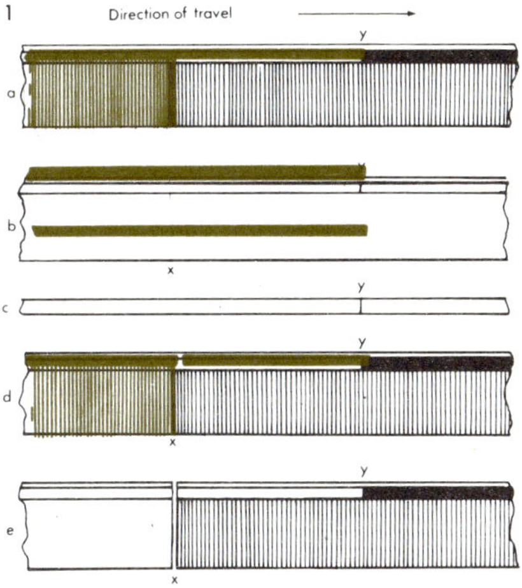

1 Cut and join
In quadruplex videotape recording the sound and vision recordings do not coincide. Sound precedes vision by about 23 cm., or 15 picture frames. a, end of first sequence. b, sequence to be added. x, vision cue. y, sound cue. To make a continuous sound edit, first the vision cues are marked on the tape opposite the vision head with a felt pen. The sound from the start of tape b is copied on the the audio tape c. The audio and video tapes are set together at a point about 10 sec before the cue and run together. As the cue mark passes over the record head the VT is switched to record sound only from the audio tape. The start of the sound from the next sequence is now tacked on to the end of the previous one (d). When the tapes are joined at the cue mark the sound and vision should match.

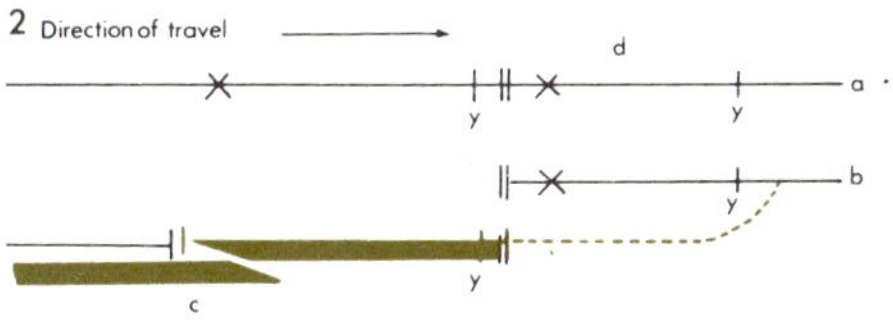

2 Roll-back and mix
a, recording on main recording tape. b, backing copy. To mix to a new sequence at x which has been duplicated on both machines, point y is located prior to the cue at which a butt join can be made later. The backing machine is then wound back to about 10 sec. before this point and put into repro mode. Machine a then records from machine b through the studio sound and vision mixers. When the cue point is reached the studio is cued and a mix to the studio accomplished. Shortly afterwards machine b is put into record and continues to provide a backing copy. Subsequently the tape will be cut to remove the duplicated section d on tape a.

Sound and Electronic Editing

The cutting of tape is being superseded by electronic editing. For this purpose a VTR machine fitted with an electronic editor is required. This is a device that controls the switching logic to make allowance for the distance between the erase and record heads so that the machine can be switched instantly from replay to record, or vice versa, without losing signal continuity.

Studio editing
The studio editing process is as follows: At the 'out' cue point at the end of a studio sequence the director presses a button which records a cue pulse on the tape and the machine is stopped. The next sequence is then set up and the machine wound back about 20 seconds. The VTR machine is then restarted and, if required, the sound fed back to the studio to help the actors to pick up their cues. About 1·3 sec. before the editing point a short burst of tone known as the 'advance cue' is relayed by a loudspeaker in the production control room. It can be heard by everybody listening to talkback as a warning that the cue is coming. When the cue pulse is reached the studio monitors automatically switch in vision and sound from the VTR to the studio output. This enables the director to judge the effectiveness of the edit. If it is not correct the cue pulse can be re-recorded or, if the error is slight, the VT switch/pulse time relationship can be altered by up to 12 or 18 frames (according to machine type). When the correct edit point is established the process is repeated in the record mode and this time the VT recorder switches to record when the cue pulse occurs.

Insert mode
The above operation can be carried out in vision or sound separately, so that it would be possible to insert a new piece of sound track or change the vision while allowing the original sound to remain.

Dub editing
Most electronic editing concerns the assembly of pre-recorded tapes by dubbing from one machine to another. It is possible to make minor adjustments to the sound in the process, but for large alterations, e.g. the addition of incidental music, the sound track is dubbed separately after the vision has been edited. This could involve the use of a multi-track audio recorder, on one track of which an address code is recorded to synchronise with an identical code on the cue track of the VTR. A copy can also be made on a less expensive helical-scan VTR (also address-code synchronised) and all operations conducted, using this as a vision reference, before finally dubbing the sound back to the quadruplex VTR.

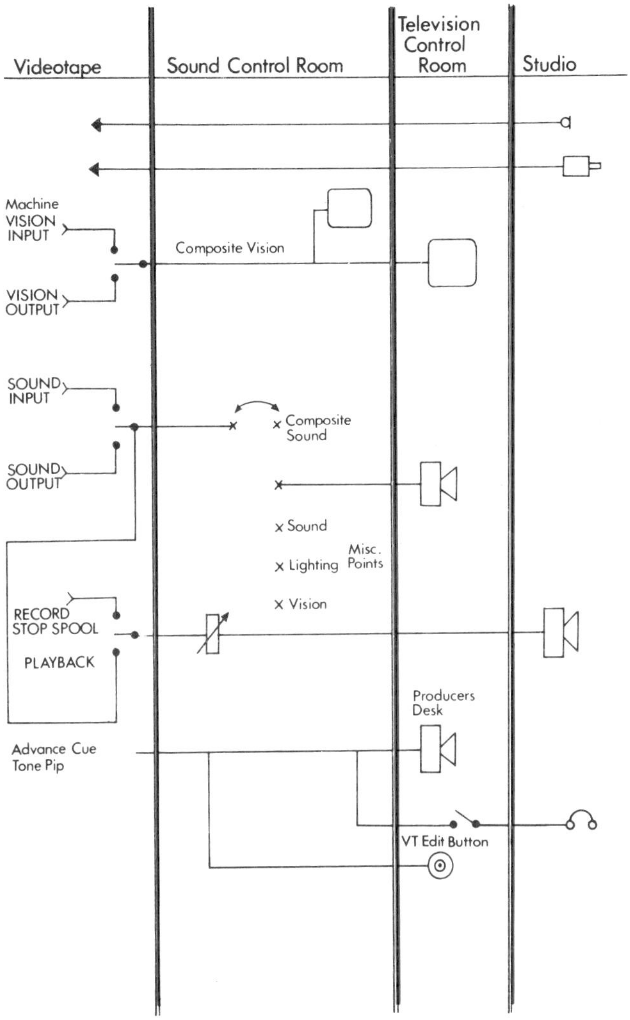

SOUND AND ELECTRONIC EDITING

Interconnections for studio electronic editing
The switches (which are in fact a ganged relay) are operated by the edit cue recorded on the tape. The up position is the record mode and down the playback mode. The monitors show either the input to the VT machine or its output according to the position of the switch. The studio loudspeaker is effective only on replay. The advance cue pip is fed to the producer's desk loudspeaker and also, if requested, to all people listening to talkback.

Dolby Noise Reduction System

Many of the processes of television sound operations involve sound recording and reproduction with the attendant risk of added system noise this can incur. This is especially true in the case of multi tracking and dubbing (page 128) and quadruplex VTR recording (page 160) which uses a narrow sound track and tape which has transversely orientated oxide particles for the benefit of the vision track which is laid laterally. The signal to noise ratio can be improved by a process of 'companding' i.e. compressing the signal volume variations before recording and expanding in complementary manner on reproduction.

A system of companding has been developed by Dolby. There are in fact basically two systems, Dolby A which is a complex system used for professional purposes.and Dolby B, a simpler system in which the compression characteristic is applied (possibly professionally) in the recording process and the expansion characteristic can be incorporated in domestic reproducing equipment.

Dolby 'A' system

There are two main problems in designing a companding system that does not cause more trouble than it cures. One is the difficulty of making the compression and expansion systems 'track' perfectly over the whole frequency and volume range. The other is preventing the background noise level going up and down with the signal, which only makes it more obvious.

The Dolby 'A' system overcomes this problem in the following ways:

The signal is provided with two parallel paths, one via a linear amplifier and the other via a differential network — the output of which is added to the 'straight through' signal when recording and subtracted when reproducing.

The differential network divides the frequency spectrum into four bands and treats each one separately so that the action is applied only where it is needed. With music for instance most of the sound energy tends to be concentrated in the lower middle band whereas tape hiss is largely high and too far removed in frequency for 'masking' to take effect. This is the effect whereby the ear tends to be deaf to sounds in the presence of, and for a short time after, hearing louder sounds of similar frequency. The Dolby system exploits this effect by restricting the action to narrow frequency bands.

The system is adjusted so that with a low level input (below −40dB) the output is 10dB higher than the input for frequencies up to about 5kHz above which it increases progressively to 15dB gain at 15 kHz. As the input signal rises above −40dB it is progressively less affected. At high levels (where the possible ill effects of companding are most obvious) the compander has least effect and is virtually bypassed.

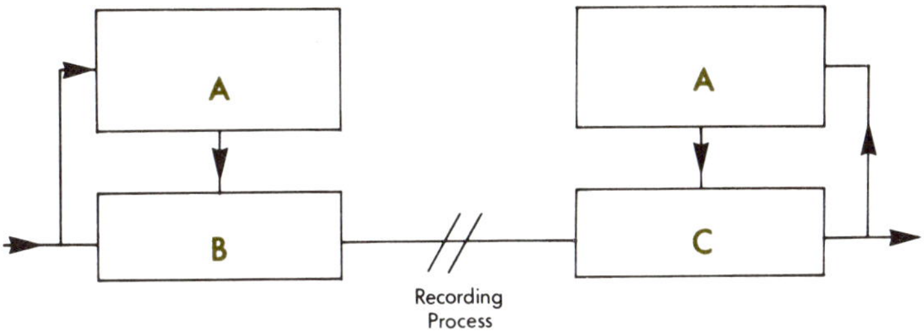

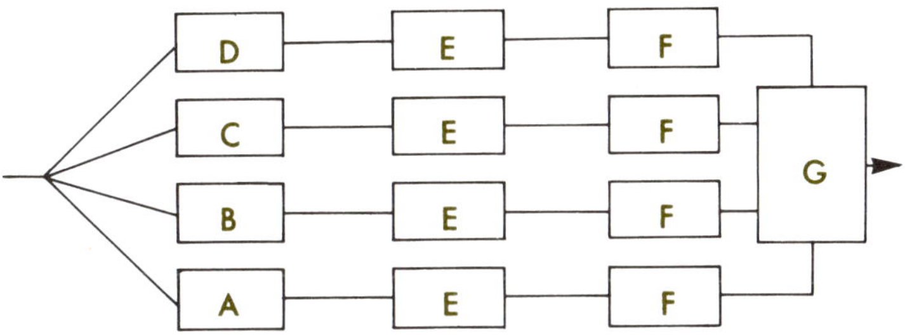

COMPANDING (DOLBY)

Companding the recording process
A, differential network. B, adder. C, subtractor. The output of the differential network is added for recording and then subtracted when replaying.

Schematic detail of differential network
A, 80 Hz low pass filter to deal with hum and rumble. B, 80-3000 Hz band pass filter to deal with mid band noise and print through. C, 3000 Hz high pass filter to suppress hiss and modulation noise. D, 9000 Hz high pass filter to suppress hiss and modulation noise. E, linear limiters. F, non-linear limiters. G, adder.

In Conclusion

To sum up we can define the main functions of the sound element in television as:

1. To provide much of the substance of the programme in terms of information and entertainment value, in an artistic and technically satisfactory manner.

2. To increase the realism and effectiveness of the pictures by simulating the appropriate acoustic conditions to suit the apparent environment and by creating the illusion of solidity of the settings.

3. To suggest movement and the third dimension by matching sound perspective to the picture viewpoint. In television particularly, owing to the small size of the screen, distance tends to be depicted in depth rather than in width. The ability to convey a sense of distance and relative perspective is an important property of sound. If the sound perspective matches the picture it will underline the realism. If it disagrees it will undermine it.

4. To enhance the mood and impact of the production with suitable music and sound effects.

The sound operation

The process of sound operations involves selecting, controlling and blending sound sources i.e. the exercise of choice on behalf of the listener. In television this must be achieved with the minimum of restriction to the visual presentation, which involves considerable flexibility of approach. No single system of sound pickup can apply to all circumstances. Each successive problem must be considered separately and the most suitable method applied individually. Full use should be made, as necessary, of sound processing techniques such as frequency response control, automatic volume compression and the addition of artificial reverberation, etc. Effective, creative sound requires imaginative forethought at the planning stage, based upon a thorough knowledge of the problems and possibilities of the medium, followed by judgment in selecting and controlling the right equipment.

The effect of the picture

The presence of a picture does not detract from the importance of the sound, or make it less critical, it merely presents a new dimension and a greater challenge.

It is worth taking care.

Further Reading

MANKOVSKY, V S:
Acoustics of Studio and Auditoria, Focal Press, London, and Hastings House, New York, 1971.
HADDEN, H. BURREL:
High Quality Sound Production and Reproduction, Iliffe, London, 1962.
A BBC manual of operational practice for sound, studio and radio outside broadcasting.
MILLERSON, GERALD:
The technique of Television Production, Focal Press, London, and Hastings House, New York, 9th ed. 1972.
A comprehensive study of the technique and aesthetics of television production, sound and vision.
ROBERTSON, A. E:
Microphones, Iliffe, London, 1963.
A detailed description of the theory of microphones with details of the design features of typical types used for broadcast purposes.
ALKIN, E. G. M:
Sounds with Vision, Buterworths, London, 1963.
A BBC manual of operational practice for television sound operators. A comprehensive study of the technique of television sound production.
GAYFORD, M. L:
Acoustical Techniques and Transducers, MacDonald & Evans, London, 1961.
A comprehensive survey of the design features of various types of microphones and loudspeakers used in broadcasting. Also included is a useful section on the acoustics of studios and listening rooms.
ORINGEL, ROBERT:
Audio Control Handbook, Hastings House, New York, 4th ed., 1972.
Elementary introduction with strong emphasis on American equipment. Covers radio and TV broadcasting.
NISBETT, ALEC:
The technique of the Sound Studio, Focal Press, London, and Hastings House, New York, 3rd ed. 1974.
Full treatment of the subject including stereo techniques, sound effects, programme construction, radiophonic techniques, tape editing, etc. Covers radio and TV.
NISBETT, ALEC:
The Use of Microphones, Focal Press, London, and Hastings House, New York, 1974.
Description of microphone types, their uses, positioning, and balance for microphonic radio and television broadcasts.

Glossary

Acoustic Effect (68) The effect of the surrounding environment on the sound.

Acoustic Instrument (152) A musical instrument in which the sound is produced naturally i.e. without electronic equipment.

Acoustic Treatment (70) Material applied to the surface of an enclosure to modify its acoustic properties.

Ambient Sound (10) The noise, reverberation or atmospheric sounds that form a background to the principal source.

Ambiophony (74) A method of adjusting the reverberation within a studio or concert hall, by means of a loudspeaker feedback system in which time delays are incorporated. Loudspeakers mounted around the walls and ceiling simulate natural sound reflections that would have occurred if the surfaces had been reflective instead of absorbent.

Animation (162) The technique of building up a moving picture sequence by joining together small sections or single frames.

Artificial Reverberation (72) Simulation of the natural effect created by multiple sound reflections within an enclosed area.

Atmosphere (124) Background sound or environmental effect.

Attack Time (64) The time required for the action to take effect in an automatic volume compressor or limiter, following the sudden application of a signal larger than the limiting (or onset) value.

Audience Reinforcement (PA) (86) The relaying of programme sound to the audience present at a production. The term 'reinforcement' (as opposed to Public Address) suggests that the live sound would be audible to the audience, albeit at low volume, without the use of loudspeakers but this is seldom the case in modern vocal presentation.

Aural Perspective (96) The impression of distance of sound source created by such factors as relative volume, proportion of direct to reverberant sound and characteristic quality etc.

Backing (Vision) (110) Background scenery.

Backing (Sound) (154) accompaniment.

Backing (Tape) (128) The base (e.g. acetate, PVC which carries the oxide coating for recording.

Backing Copy (162) a duplicate recording made as an insurance against technical faults occurring on the main copy.

Backing Track (158) Pre-recorded accompaniment to which the vocal or solo part can be added.

Back Tracking (158) A method of building up a sound recording in which the artists add successive parts while listening to their previous recording.

Back Projection (110) a method of providing scenic backgrounds by projecting a still or moving image on to a translucent screen behind the action.

Balance (14) The selection of microphone positions and volume levels from a number of sources to promote an artistically correct effect.

Boom (36) A mobile device incorporating a telescopic arm for varying the position and direction of a microphone to follow mobile action.

Cartridge (128) In current practice this term is used to describe a box containing a single spool which dispenses a continuous loop of tape. The tape, which is lubricated to reduce friction, feeds from the inside of the spool and is returned to the outside. The spool is not driven, movement of the tape being caused entirely by the capstan and pinch wheel. Normally 6.3 mm tape is used, running at 9.5 cm/sec. Eight tracks are available as four stereo pairs.

Cassette (128) A reel-to-reel tape system in which both the feed and the take-up spool are incorporated in a plastic box. The tape, approx 3.5 mm wide is driven at 4.75 cm/sec. Four tracks are available for stereo operation, one pair used in each direction, changed by turning the cassette over. As the stereo pair are together, occupying half the tape width, mono and stereo recordings are compatible.

Cello Board (138) A board designed to accommodate the chair of a cellist and the spike of the cello, to provide a stable base with good acoustic properties for the instrument.

Clean Feed (92) An output feed to a contributor which contains all the contributions except his own.

Clean Feed Talkback (52) A circuit which normally carries clean feed but can be intercepted by a talkback microphone and switch.

Closed Circuit A programme not intended for transmission to the public.

Commag (120) film with combined magnetic sound.

Comopt (120) Film with combined optical sound.

Companding (166) A method of applying volume compression to a signal before a process (e.g. recording or transmission) and then corresponding expansion afterwards.

Composite Sound Complete programme sound i.e. commentary with international sound.

Compressor (64) A device for reducing automatically the dynamic range of a signal.

Conference Network Circuits connecting a number of remote points to provide simultaneous speech to or from any points.

Crab (40) To move sideways.

Cue Programme A circuit carrying audio programme material, used for cueing or monitoring purposes.

Cyclorama (118) A plain curved background normally made of stretched fabric.

Dead Theatrical term meaning not required further for the production.

Decibel (10) A measure of relative intensity used for sound measurements because it obeys a logarithmic law as does the human ear in relating sound intensity to sensation. One decibel represents about the minimum change of level that can be appreciated by the ear and that only on a steady tone. On programme a change of 2 dB is just about discernable.

The gain of amplifiers is usually expressed in dB(power) = 10 log P1/P2 where P1 and P2 are the input and output powers, or as a voltage gain in dB (voltage) = 20 log V1/V2 where V1 and V2 are the input and output voltages measured across equal impedances. The voltage dB is twice the power dB for a given ratio, the power ratio in a circuit being proportional to the square of the voltage ratio. Hence a 2:1 power ratio would be 3 dB; as a voltage ratio it is 6 dB.

Diffusion (68) The random nature of the reflective paths for sound within an enclosure.

Diversity (28) A reception system embodying a number of aerials offering the possibility of selecting the best signal from instant to instant.

Dolby (166) A method of improving the signal/noise ratio of a recording or transmission system by means of a frequency-selective volume companding (compressing and then expanding) process. There are basically two alternative systems. The Dolby A System is used professionally and divides the frequency spectrum into four bands. Each band is provided with its own compressor (and later expander) so that the action is applied only where it is needed. It thus eliminates the effect of controlling sounds of one frequency by another and makes full use of the masking effect of the ear. The Dolby B System is mainly used for commercial recordings (particularly cassettes). The process is applied to the recording and a corresponding process is incorporated in the reproducer. The signal passes through a 19kHz low pass filter and then divides, one path going direct to an adder and the other going to the adder via a variable high-pass filter and an amplifier. The cut-off frequency of the high-pass filter is controlled by the signal level so that it reduces as the level is reduced. When the input to the equipment is high enough to mask the h.f. noise the filter passes only extreme high frequencies and the effect on the output is small, most of the signal being supplied by the direct path.

Dubbing (164) The process of re-recording from one recording to another. The term is also used for the addition of sound, e.g. dialogue to a previously recorded picture.

Dubbing (Post) (162) The addition of material, or a complete sound track, to accompany pictures shot previously.

Dynamic Range (10) The ratio (in phons) between the softest and loudest sound. Or the ratio (in decibels) of signal variation available between overload and unacceptable signal/noise ratio.

Echo (68) Discrete, separately identifiable, repetitions of a sound due to reflections from hard surfaces. This term is sometimes used erroneously to represent reverberation.

Electret Diaphragm (16) A diaphragm used for electrostatic microphones which has been given a permanent electrostatic charge, thereby eliminating the need for polarising supplies.

Electrostatic Microphone (16) A microphone that operates by virtue of variations in capacitance between the diaphragm and the back-plate spaced closelv behind it.

Erasure (128) The process of removing previously recorded signals from magnetic materials prior to recording.

Extinction Frequency (160) The frequency at which signal cancellation occurs in magnetic recording owing to the width of the reproducing head gap being equal to the recorded wavelength. The extinction frequency defines the upper frequency limit of the direct magnetic recording process. The lower frequency limit is also a function of head-gap size. The output is proportional to rate of change of magnetic flux across the gap which reduces (at the rate of 6 dB per octave) as frequency is reduced and the wavelength becomes long in relation to the head gap. The normal maximum range for direct magnetic recording is therefore limited to about 10 octaves.

Eye Line The direction in which artists look while performing.

Feedback (84) A proportion of the output of an amplifying system that is returned to the input.

Fishing Rod (30) A short pole supporting a microphone carried by an operator, used to pick up mobile sources of sound in cramped situations or outside locations where a boom would be inappropriate.

Flat (94) a piece of hard scenery.

Foldback (60) A source of sound such as music, effects, or remote participant relayed into the studio by loudspeaker (sometimes mobile loudspeakers) for the benefit of the artists.

Frequency Modulation (160) A method of modulating a carrier by varying its frequency as opposed to its' amplitude.

Gain Amplfication of an amplifier or component usually expressed in decibels.

Guide-Track (128) a specially recorded sound track (normally one track of a twin-track recording) used for synchronising purposes. If, for example, an orchestral accompaniment is to be added live to a pre-recorded vocal, the guide track, possibly containing a bar count, is fed to the conductor on headphones to keep the orchestra in time during an introduction or non-vocal reprise.

Hertz (Hz) (10) Frequency, cycles of repetition per second.

Howlround (84) Instability in a loop usually including an electrical and an acoustic path, e.g. loudspeaker and microphone, where the electrical amplification is greater than the acoustic loss. If the gain is progressively increased the initial result is an uneven increase in response due to positive feedback peaking at the frequency at which the electro/acoustic circuit is most sensitive, finally building up into a continuous howl at this frequency.

Indirect Sound (68) Sound which reaches the listener, or microphone, by acoustic reflection.

Intensity of Sound (10) The intensity is a measure of the power of a sound. It is usually measured in decibels relative to the threshold of hearing at 1000 Hz. This can be represented by a pressure range of 1/5,000 bar or 0·0002 dyne/cm^2 at the ear of the listener. The decibel scale is used because it is logarithmic and the ear has a logarithmic response. One dB represents roughly the smallest perceptible change of sound intensity.

Sound intensity is not the same as loudness (which is measured in phons) except at 1000 Hz because the ear has an unequal frequency response, which also varies with intensity (see **Phon**).

International Sound (92) Feed of clean effects (without commentary) to be fed to several countries for the addition of commentaries in their own languages.

Lazy Arm (24) A simple form of boom or counterweighted arm mounted on a stand that can be used for suspending a microphone over an object such as a music stand or piano.

Leader (130) A section of uncoated tape, normally coloured white, which is joined to the beginning of a recording to allow for threading to the take-up spool and on which details of contents can be written.

Limiter (64) A device for automatically preventing the volume of a signal from rising above a pre-set value, thereby preventing overload. It usually takes the form of an amplifier with zero gain (i.e. the signal comes out at the same value as it goes in) until the pre set limiting value, known as the onset point, is reached. Above this value the amplifier has greatly reduced gain so that a very large increase in input signal results in an insignificant increase in output. The action is similar to a compressor except that the grain reduction is more severe (a compression ratio in excess of 10:1 can be considered as limiting) and the time constants are different. A limiter used to protect against overload (e.g. in an AM transmitter) requires a very rapid attack time and relatively slow recovery time, i.e. it reduces the level instantly and restores it slowly.

Line-up (56) The adjustment of levels in a piece of apparatus or programme chain.

Mute (Film) Silent film.

Mute (Instrument) (142) Device for reducing the tone and altering the characteristic quality of an instrument.

Noise (10) Unwanted sound. The lower limit of dynamic range of a sound reproducing system is set by the requirement to provide a reasonable separation from the background noise level. This might be due to extraneous sounds, amplifier hiss, ventilation rumble, or line noise etc. Noise which contains an equal value of power over the whole audio frequency range is known as 'white noise'.

Obstacle Effect (16) Sound waves are able to curve around obstacles whose dimensions are smaller than their wavelength. This effect also depends on the shape, i.e. the streamlining of the object.

Optical Sound A film with a sound track which takes the form of variations in light transmission. It is reproduced by shining a light source through the optical track on to a photo electric cell.

Overtones (10) Harmonics. Frequencies which extend in multiples of the fundamental note and give a tone its characteristic quality.

Pan To change direction in the lateral sense.

Phase (10) the position in the cycle that a waveform has reached at any given instant. Waves are said to be in phase when their position in the cycle coincides.

Perspective (96) The apparent effect of distance. The distance of a sound source is suggested by the relative volume level, the ratio of direct to reverberant sound and the characteristic quality.

Phon (10) A measure of the loudness of a sound that takes into account the unequal frequency response of our ears. Phons and decibels are the same at 1000 Hz. At other frequencies their relationship is illustrated by the curves of equal loudness (page11).

Pitch (10) Subjective effect of sound related mainly to frequency but affected also by intensity and harmonic structure. As volume is increased, high sounds can seem higher and low sounds lower.

Practical An object, such as a scenic effect or property, that works, i.e. performs its proper function.

Pre-hear (60) A circuit, normally controlled by a key, by which it is possible to listen to the input to a fader or control before it is faded up.

Presence Filter (66) A filter which imparts a rise, usually in the form of a steep peak, to the frequency response of a circuit in the region of 3-7 kHz. The frequency range in which the sibilants in speech tend to lie, thus making for clarity and the effect of closeness or 'presence'.

Prompt Cut (60) A device used in association with live drama. It consists of a button, carried by the prompter, which when pressed cuts the output of the studio microphones, (but not the grams or tapes) so that a prompt can be given to the artist.

Property An object used to dress a set or carried by an artist.

Pulse Code Modulation A modulation system in which the signal carrier wave is broken up into a series of pulses. The signal information can be contained in the position of the pulses (relative to a marker pulse) the width (mark space ratio) or their amplitude in relation to their position. These are translated into a series of numerical values which describe the waveform from which they were derived. PCM is used for the transmission of sound over landlines and radio links. The sound accompanying a television picture can be coded and inserted as a series of pulses into the synchronising signal between each television line. A 10 digit code is used. In a time-division-multiplex system it is necessary to sample at twice the highest frequency it is required to transmit. For a 625 line television system this is conveniently 31.25 kHz, twice line frequency, which allows an audio bandwidth of 14 kHz. This is achieved by delaying alternate sets of pulses by half a line duration, inverting them and then inter-leaving them with the next set of pulses, representing the next sample. Thereby inserting all 20, plus a marker pulse, in each sync pulse. The type of code used is binary, each pulse being given a value of 1 for 100% modulation and 0 for zero. The system is thus very resistant to amplitude distortion as the receiving equipment only has to recognise all above 50% modulation as 1 and all below as 0. Pulse code systems tend to provide excellent quality over long lines and radio links without progressive deterioration (as in analogue systems) until the point where they fail completely.

Radio Microphone (28) A microphone (usually a lavalier or hand-held type) equipped with a miniature transmitter to enable the artist to carry it about without the encumbrance of a cable.

Reverberation (68) The sustaining effect of multiple sound reflections within an enclosure.

Reverberation Time (68) The time taken for a sound to die away through 60 dB (one millionth of its original intensity) in a reverberant enclosure after it has been cut off.

Roll-Back and Mix (162) The process of repeating, by means of a recording, the end of a previous sequence in order that it can be mixed into the start of the next sequence.

Sound Image (96) The mental picture created by a particular character of sound.

Spot Effects (124) A sound effect made by acoustic means (as opposed to a recorded effect).

Straight Orchestra (144) An orchestra so constituted and playing music so orchestrated that it produces a balanced sound without the aid of microphones.

Strike To remove scenery or an object from a stage or setting.

Talkback (52) Open microphone to one or more destinations. Used for cueing and briefing instructions.

Talkback (Lazy) A communication system often used at outside broadcast sites in which a microphone at the contributor's position (not the broadcast one) is connected continuously to enable him to speak to the production area.

Talkback (Lazy - Clean Feed) As above except that the broadcast microphone is used, connected at a point before the fader.

Telecine A machine for producing Television pictures from motion picture film.

Tremolo A rhythmic variation of volume level at subsonic frequency i.e. between about 5 to 25 Hz causing a type of 'flutter' effect.

Unidirectional Microphone (18) A microphone that picks up sound essentially from one direction only.

Vibrato A rapid variation in pitch used by singers, string players, trombonists and in some electric organs to enhance the quality of sustained notes.

Volume Unit Meter (62) A meter which indicates programme volume in volume units. i.e. percentage utilisation of the channel.

Wild Track (120) Sound recording intended to accompany a picture but not synchronised to it.

Windshield (38) A device fitted around a microphone to protect it from wind or air turbulence due to rapid movement through the air. e.g. on the end of a boom. Windshields have to be large to be effective against wind. Smaller *close-talking shields* are available for some microphones to guard against breath puffs.

Wow A variation of pitch, normally most noticeable on low notes in recordings, due to speed instability in the recording or reproducing mechanism.

Zero Level The level used for lining up equipment and the broadcasting chain using standard tone. It corresponds to a power of 1 milliwatt in 600 ohms and is indicated by the figure 4 on a peak programme meter.